Future of Business and Finance

The Future of Business and Finance book series features professional works aimed at defining, analyzing, and charting the future trends in these fields. The focus is mainly on strategic directions, technological advances, challenges and solutions which may affect the way we do business tomorrow, including the future of sustainability and governance practices. Mainly written by practitioners, consultants and academic thinkers, the books are intended to spark and inform further discussions and developments.

Tamara Carleton • Shaun West
William R. Cockayne
Editors

Inventing the Almost Impossible

Creating, Teaching, Funding, and Leading Radical Innovation

Springer

Editors
Tamara Carleton
Department of Mechanical Engineering
Blekinge Institute of Technology
Karlskrona, Sweden

William R. Cockayne
Institute for Information Management
University of St. Gallen
St. Gallen, Switzerland

Shaun West
Institute of Innovation and Technology Management
Lucerne University of Applied Sciences and Art
Horw, Luzern, Switzerland

ISSN 2662-2467 ISSN 2662-2475 (electronic)
Future of Business and Finance
ISBN 978-3-031-36223-1 ISBN 978-3-031-36224-8 (eBook)
https://doi.org/10.1007/978-3-031-36224-8

This Springer imprint is published by the registered company Springer Nature Switzerland AG
The registered company address is: Gewerbestrasse 11, 6330 Cham, Switzerland

Contents

We Need a Discipline of Invention

Tamara Carleton, William R. Cockayne, and Shaun West

Abstract This chapter introduces the book *Inventing the Almost Impossible*, which combines various case studies, examples, and lessons for creating, teaching, funding, and leading radical innovation. The book's broader aim is to establish a scholarly foundation for the study of invention, which is a fundamental human capacity that has shaped society since prehistoric times, yet it has not been given the full attention it deserves. We have invited a spectrum of authors from academia, industry, and other areas of work to consider an angle for inventing almost impossible ideas, which are just within the realm of possibility and have the potential to produce radical innovation. Overall, this collection seeks to raise questions that explore the multifaceted aspects of invention more deeply and establish invention as a complementary element of technological creation, research into development, and new product development.

Why doesn't a proper discipline of invention exist? Science historian James Burke—some readers may recall the seminal BBC television show *Connections* from the late 1970s that he created and hosted—describes invention as the "act of bringing ideas or objects together in a novel way to create something that did not exist before" (Britannica, 2023, par. 1). In the *Encyclopedia Britannica* under the invention entry, Burke notes that "Ever since the first prehistoric stone tools, humans have lived in a world shaped by invention. Indeed, the brain appears to be a natural inventor" (ibid.,

T. Carleton (✉)
Department of Mechanical Engineering, Blekinge Institute of Technology, Karlskrona, Sweden
e-mail: tamara.carleton@bth.se

W. R. Cockayne
Institute for Information Management, University of St. Gallen, St. Gallen, Switzerland
e-mail: william.cockayne@unisg.ch

S. West
Institute of Innovation and Technology Management, Lucerne University of Applied Sciences and Art, Horw, Switzerland
e-mail: shaun.west@hslu.ch

T. Carleton et al. (eds.), *Inventing the Almost Impossible*, Future of Business and Finance, https://doi.org/10.1007/978-3-031-36224-8_1

para. 2). Yet the study, practices, and even teaching of this human capacity for creation are too often dismissed as corporate research or entrepreneurship activity, tracked as late-stage patent output, or treated as a crackpot hobby.

The process of inventing big ideas, which lead up to and can produce radical innovation, is even less understood or researched—even though multiple groups from industry (e.g., Tesla) to government (e.g., NASA) to defense (e.g., DARPA) regularly pursue, fund, and aim to commercialize game-changing inventions (Carleton & Cockayne, 2023). Given the outsized impact of inventions in business and society across history, we feel that the topic of inventing the almost impossible merits serious study.

We chose to describe the nature of these ideas as *almost impossible* because we want to bring the event horizon of reality in closer. These ideas are just within the reach of possibility, so they should not strike certain critics as creeping into the preposterous or the realm of science fiction. That said, when we spoke with Rick Satava, a seasoned DARPA program manager and the father of the field of telerobotic surgery, he immediately said, "I don't think you need 'almost impossible' in your book title, mainly because what DARPA does is to fund the invention of things people claim are impossible" (personal communication, February 2, 2023). Similarly, Alan Kay—who was seminal in imagining and inventing the foundation of the modern user-centric personal computer, among other work—explained in a Stanford University lecture that ideas feel impossible due to thinking largely in today's terms. Kay said:

> The problem with the present is [that] is so glittery and distracting; there's so much stuff going on. It's hard to think about anything except the present. And if you're thinking about the present, your ideas are going to be derived from the present. And therefore, they're going to be incremental. (Stanford Online, 2017, 2:13)

Consider the original moonshot as an example. When US President John F. Kennedy declared to the American public in 1962 that, by the end of the decade, the United States would land astronauts on the moon, he knew that many would see the vision as audacious and even impossible. No country had achieved this milestone, and at the time of this book's publication, only four countries have landed on the lunar surface. While hard, Kennedy felt the vision was possible within the 10-year timeline. More importantly, his administration recognized the spillover effects that this effort would catalyze. Often forgotten are Kennedy's follow-on remarks in the same speech. He stated, "The growth of our science and education will be enriched by new knowledge of our universe and environment, by new techniques of learning and mapping and observation, by new tools and computers for industry, medicine, the home as well as the school. Technical institutions, such as Rice [University, where he gave the speech—editors], will reap the harvest of these gains. And finally, the space effort itself, while still in its infancy, has already created a great number of new companies, and tens of thousands of new jobs" (Kennedy, 1962, para. 23–24).

To get to the moon, new technologies would need to be invented to achieve what many saw as impossible. And over 60 years later, humanity is still benefiting from the various technologies, businesses, tools, and even ways of working that were invented because of the space programs. Beyond aerospace, more knowledge and techniques for inventing the almost impossible are needed today to overcome multiple grand challenges facing communities and nations worldwide, such as feeding 10 billion people, keeping global heating below 1.5 °C, providing healthcare to aging populations, addressing the next waves of global pandemics, and more.

In this book collection, a motley collection of papers aims to spark the creation of a scholarly foundation for studying invention. Alongside the chapter authors, we seek to raise the next set of questions to explore the multifaceted aspects of invention more deeply. A new research agenda is needed so that we can understand better what works (and why) alongside what doesn't work (and why) in invention. The research focus should deeply study how teams lead the transformation of strategic imagination into radically new ideas and disruptive solutions, along with investigating the underlying cultural dynamics and decision-making models. Broadly speaking, whereas studies of innovation and entrepreneurship address new venture creation and commercialization, and studies of creativity address elements of idea generation and novelty, then studies of invention would extend this research spectrum by focusing on complementary elements of technological creation and research into development, spilling over into new product development. In terms of researching the education and capacity development for invention, students of all ages would benefit from increasing their fluency and skills for seeing what does not yet exist and ways for gaining an inventive ability. Notably, how can universities and teachers better educate, coach, train, and empower the youth who will invent and shape the future?

As editors of this collection, we have drawn on our combined breadth of experiences across multiple roles in industry and academia. As university researchers, we have investigated and documented the best practices of pioneering radically innovative groups like DARPA and ARPA-E. As professors and instructors, we have taught and overseen numerous courses and programs—such as the Silicon Valley Innovation Academy at Stanford University—that seek to build invention capacity in teams and organizations. As practitioners, we have introduced new methods and tools to support radical innovation, patented our own inventions, and converted some of these inventions into new businesses. And as instigators, we have convened seminars and conferences to spark new thinking and new connections between communities, which helps to create the underlying conditions that let an inventive and innovative culture flourish.

Based on these experiences, we see how much more is needed to build a new field of study. We intend for this book to serve as an opening to the broader theme of invention. So, who should read this book? We deliberately expect a mixed audience. Those actively inventing or leading radical innovation will find certain chapters helpful in guiding their work. People who teach innovation or creativity should read this book because the case studies will show more ways of how invention happens in the wild and prompt new questions for scholarly inquiry. Funders of innovation need to read this book. They will discover some lessons that confirm how multi-helix collaboration and unconventional partnering can yield amazing outcomes. Overall, our book describes real stories and examples of invention as a key part of the journey from imagination to innovation through different lenses, which we hope offers inspiration to everyone teaching, funding, and leading at the edge of creating new fields.

Instead of insisting on the same chapter formula, we instead encouraged authors to present their topic as it made the most sense. The unifying objective is to help readers understand the current state of knowledge related to invention, including some of the thorny problems when teaching, funding, or leading inventive work. We hope that the loose, essay-style approach for this book offers more accessible short stories and diverse case studies, which in turn uncover multiple insights and hints on what a broader research agenda should consider—and perhaps where you can take it next in your own work. In addition, we deliberately broadened the readership to reach others outside academia, as these voices are needed to help inform a new field of study and provide a source of raw evidence to spark new theories and models.

Where does this dialogue go next? Those inventing—the practitioners—are typically too busy to pause and document their knowledge, let alone aggregate their lessons with others outside their circles. Those in academia can help capture this knowledge and further build a shared knowledge base of new studies with models and theories that become the basis for a proper field of study. This collection is one stepping stone toward this goal, uniting one set of diverse voices of practitioners and scholars. We need more stones to build a foundation. As more people understand how to invent the (almost) impossible, more groups can better address society's pressing challenges and think up new types of solutions that do not yet exist. This means we need to understand the best of today so that we can imagine new ways to create the future.

References

Burke, J. (2023, April 5). *Invention*. https://www.britannica.com/technology/invention-technology

Carleton, T., & Cockayne, W. R. (2023). *Building Moonshots: 50+ ways to turn radical ideas into reality*. John Wiley & Sons.

Kennedy, J. F. (1962, September 12). *Address at Rice University on the nation's space effort* [Speech transcript]. John F. Kennedy Presidential Library and Museum. https://www.jfklibrary.org/learn/about-jfk/historic-speeches/address-at-rice-university-on-the-nations-space-effort

Stanford Online. (2017, May 9). *How to invent the future I. Stanford CS183F: Startup school* [Video]. YouTube. https://youtu.be/id1WShzzMCQ

Tamara Carleton is a globally recognized expert in radical innovation. As founder and CEO of Innovation Leadership Group, Tamara works with business and government leadership teams to help them build innovation capability and lay out bold futures. With a penchant for creating new methods and tools, Tamara is also a renowned educator, teaching strategic foresight and innovation at multiple universities worldwide. She holds a doctorate in mechanical engineering from Stanford University, where she explored DARPA's enduring innovation practices. She is the author of *Building Moonshots: 50+ Ways to Turn Radical Ideas into Reality* and the *Playbook for Strategic Foresight and Innovation*.

William R. Cockayne has spent his life building great teams who can imagine, invent, and deliver the future. He is a visionary technologist with a passion for understanding the latest technologies. He holds a doctorate in mechanical engineering from Stanford University and is the co-author of *Building Moonshots: 50+ Ways to Turn Radical Ideas into Reality* and the *Playbook for Strategic Foresight and Innovation*.

Shaun West gained a Ph.D. from Imperial College in London and worked for over 25 years in several businesses related to industrial services. He started his industrial career with AEA Technology before moving to National Power, where he developed and sold services to external companies. After an MBA at HEC (Paris), he moved to GE Energy Services, modeling and negotiating long-term service agreements. At Sulzer, he drafted the strategy that led to the service division tripling in size over 10 years and executed part of the strategy by acquiring a 220M CHF service business. Now at the Lucerne University of Applied Sciences and Arts, he is the Professor of Product-Service System Innovation. He focuses his research on supporting industrial firms to develop and deliver new services and service-friendly business models. He is passionate about exploring complex systems to understand problems better and build solutions faster. He is a member of the advisory board for ASAP Service Management Forum and a member of the Swiss Alliance of Data-Intensive Services. Shaun lives close to Zurich with his wife and two children. He climbs, skis, and runs.

Re:Design Thinking

Shaun West, Thomas Schönweitz, and Arman Amin

Abstract The rise of human-centered design thinking in the 1980s could not predict the growing number of products and services we would invite into our lives. As we continue to innovate and build upon the design of these devices, our emphasis on design shifts to support the weak points of these outdated systems rather than empowering the humans who inhabit them. This leads to short-term fixes rather than long-term improvements, driving a wedge further between the actors within the ecosystem and the solutions these systems are trying to solve. Short-term thinking has put us in this situation. For us to move forward as a species, we need to stop kidding ourselves and focus on improving the human condition rather than continuing to serve the same what we have designed for ourselves. Through long-term thinking, visionary movements, and building for a better tomorrow, we can create a better environment for all actors involved in the systems.

1 Innovation by Design: Reimagining Solutions for a Connected World

Our inspiration for this essay came from a simple "what if" question. "What if your specific coffee order awaited you before catching the train into the city?" This evolved into the larger question of "What if everything just worked?" These

S. West (✉)
Institute of Innovation and Technology Management, Lucerne University of Applied Sciences and Art, Horw, Switzerland
e-mail: shaun.west@hslu.ch

T. Schönweitz
Munich, Germany
e-mail: thomas.schoenweitz@whitespring.de

A. Amin
Send More Brains Inc., Toronto, Canada
e-mail: arman@sendmorebrains.com

T. Carleton et al. (eds.), *Inventing the Almost Impossible*, Future of Business and Finance, https://doi.org/10.1007/978-3-031-36224-8_2

questions allow us to explore hypothetical situations and spark our imaginations to innovate. These questions are more important than ever in our interconnected world, where people, processes, and things constantly interact.

Unfortunately, many innovations today are limited by their ecosystem and need to consider the broader context in which they exist. Innovation also needs to consider the consequences of how different systems interact to create new norms and expectations around old problems and challenges. This is a design failure, as it often feels like teams of engineers, artists, designers, and programmers conspired to create a less-than-ideal experience that punishes the individual for not choosing their specific product suite. This approach breaks down complex issues into a simplistic view of "needs" for a "user," ignoring the richness of our environment and neglecting multiple perspectives and timeframes.

In this essay, we explore these aspects and seek to answer how we can achieve better innovation, an innovation that helps both people and the planet we inhabit and share. We propose an approach allowing us to imagine, adapt, and deliver innovation considering the broader context. We can co-create value in our increasingly connected world by exploring the hype around innovation management tools, understanding the context of past thought, and considering the human and the planet within the context of future innovations.

Through this exploration, we present a RE:design of Design Thinking that helps us become better innovators, capable of designing solutions that empower users and create a better world for all.

2 Beyond Hype: The Realities of Innovation Management Tools

The world of innovation management tools is vast, with the hype cycle from Gartner providing a valuable guide to the tools available (Dedehayir & Steinert, 2016). The cycle positions various tools, with open innovation, trend management, hackathons, business models, and idea/innovation workshops now considered mature. The forefront of innovation is now centered on emerging forms of digital innovation, ISO 56000, foresight, centers of excellence, storytelling, and ecosystem innovation. Meanwhile, some commonly used approaches to innovation, such as lean start-up and design thinking, are now on a downward trajectory, suggesting the need for adaptation to remain relevant and overcome the weaknesses of these approaches.

To understand innovation, we must first define it. Innovation, derived from the Latin verb "innovāre," refers to introducing new things, ideas, or ways of doing something (Keeley et al., 2013). However, innovation is not limited to technological invention and extends across many managerial fields (Dodgson et al., 2013). Innovation requires a sustainable growth mindset focusing on the future and action, a set of trade-offs that build upon each other. However, innovation also means change, which individuals and organizations may resist for different reasons. There remains a paradox of innovation: who wants change versus who wants to change (see Fig. 1).

Fig. 1 The paradox of change associated with innovation (artwork by Maryam Afshar of "Studio401")

Innovation is a complex process with many issues that limit the predictability of its outcome. Past metrics are not guarantees of success, and any value proposition has a half-life, as confirmed by the S-shaped curves (Sawaguchi, 2011). Despite the tools available to support innovation, there are areas for improvement in the innovation management tools available for people, society, and the planet. The consequences or implications for others are often not factored in, meaning the system-of-system perspective is missed. This results in innovation being focused on a single and often ill-defined "user," which overlooks the interconnectivity of our world, the dynamics between different people, and new ways in which resources can be spontaneously recombined (see Fig. 2). Additionally, innovation requires a solid understanding of the existing problem(s) and a market that accepts the problem (Dawson & Daniel, 2010).

To succeed, innovation must embrace an ambidextrous approach that can drive change and create efficiencies equally valid for individuals and firms (De Bono, 2009). Innovation processes, such as open innovation, may positively affect overcoming preconceptions. However, missing from the discussion is how technology may change people's behaviors or expectations, changing the context within innovation. Although there have been attempts to put a repeatable process on innovation, such as Design Thinking, Outcome-Driven Innovation, and Lean-Startup, innovation remains an ongoing process. Thus, past decisions limit future options available to us (see Fig. 3).

There is a need for an optimistic, visionary mindset to help drive innovation, as it demands a focus on the future and action (Pinchot, 1985). Innovation requires a solid understanding of the existing problem(s) and a market that accepts the problem; with both, any creation is likely to be successful (Dawson & Daniel, 2010). Ulwick's ODI, Design Thinking, or the V-model from systems engineering all provide frameworks that support innovation. Lean does the same; in the Toyota A3 tool,

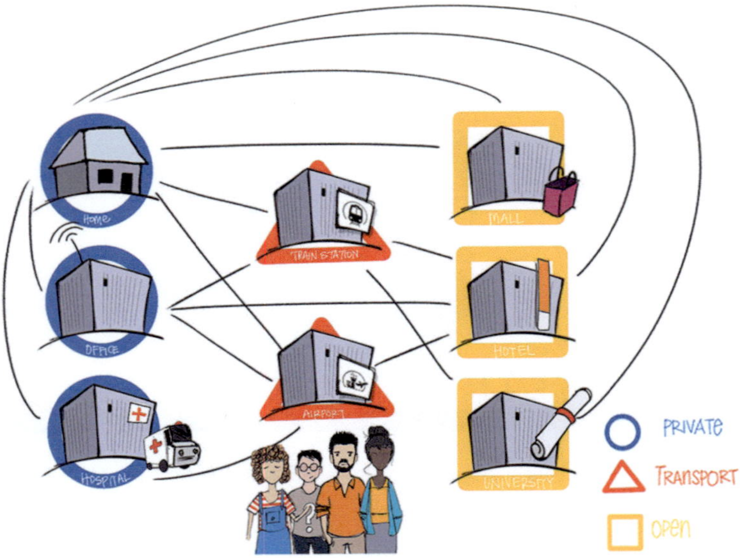

Fig. 2 Our interconnected world is a dynamic, evolving world (artwork by Maryam Afshar of "Studio401")

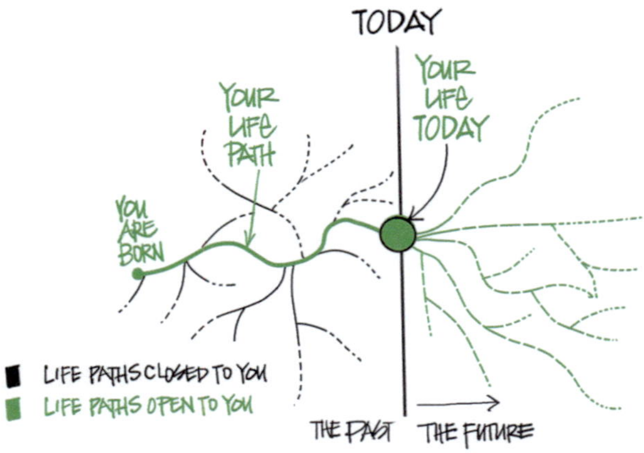

Fig. 3 Past decisions impact future options (artwork by Maryam Afshar of "Studio401")

the problem should be clear and well-defined, and a suitable solution should be built and tested. Some means should allow successful innovation; in any case, we have many examples of innovation today.

Nevertheless, there are areas for improvement in the innovation management tools for people, society, and the planet. Often the consequences or implications on others need to be factored in, meaning the system-of-systems perspective is missed. This means that innovation focuses on a single and ill-defined "user." Consequently, our world's interconnectivity is lost, and the dynamics between different people and the new ways in which resources can be spontaneously recombined are overlooked; it also misses the critical aspect that peoples' roles change depending on the given situation.

The hype cycle for innovation management tools may come and go, but the need for innovation remains constant. Tools are not a panacea; we must use them carefully and completely understand their strengths and weaknesses. Furthermore, we must recognize that innovation is not only about ideas and technology but also about people, society, and the planet. In the end, it is the human element that is the key to innovation. Therefore, we should focus on developing a forward-thinking mindset, encouraging calculated risk-taking, and fostering a culture that supports creation. Ultimately, this will lead to sustainable innovation that benefits us all.

3 The Future Is Now, "Old Man": How Understanding the Present Is Critical to Designing the Future

Innovation is a complex process that involves understanding the world we inhabit today at the macro, meso, and micro levels. However, the weaknesses of traditional innovation tools become more apparent in our increasingly interconnected and evolving world (Fig. 4). To overcome these limitations, innovation should include system design and Service-Dominant Logic aspects, focusing on the value co-creation process. In this context, innovation creates "holistic, meaning-laden experiences in nested and overlapping service ecosystems, governed and evaluated through their institutional arrangements" (Vargo & Lusch, 2016, p. 7). This approach acknowledges that innovation is a continuous process that evolves over time as actors integrate resources, provide reciprocal service, and co-create value through new value propositions (Anderson & Narus, 1998).

Apple's success in changing our expectations of what a mobile phone could do illustrates the power of visionary thinking and building outside the system. However, our understanding of the present limits our ability to design the future. As the system evolves, people's expectations change, and technology advances, our ability to anticipate the future becomes more challenging.

To design the future, we need a vision, but we must also understand what can and cannot be influenced. Actionable foresight is a form of innovation that allows us to anticipate and prepare for future events, to enable us to take proactive steps to achieve our desired outcomes. It links the past with possible futures and goes beyond mere speculation by providing possible concrete scenarios.

The service science approach to innovation with an ecosystem provides a more explicit innovation language that allows us to distinguish individuals' roles and situations. It also clearly states that only the system's beneficiaries can define value. This means that innovation changes over time as feedback loops change expectations and build new institutional arrangements. Apple was an excellent example of this, as it changed our expectations of what a mobile phone could do and catalyzed

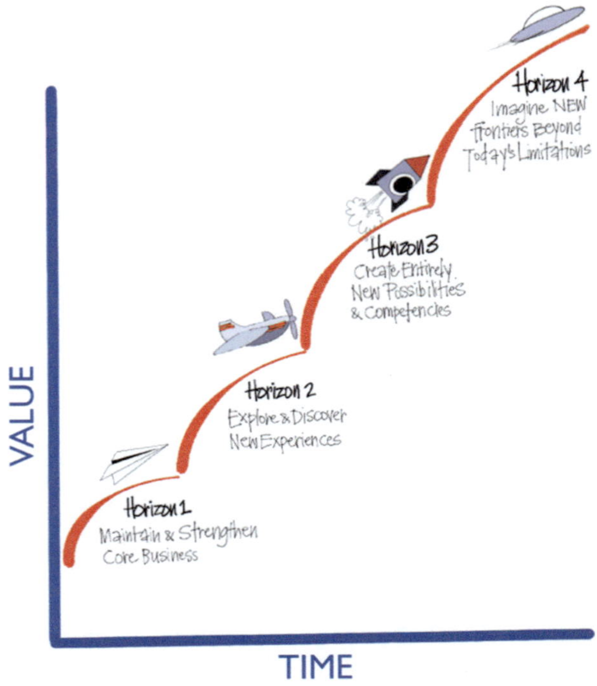

Fig. 4 Visualization of the Carleton-Cockayne model of four horizons with innovation S-curves describing long-term evolution (artwork by Maryam Afshar of 'Studio401', based on Carleton & Cockayne, 2023)

new institutional arrangements supporting value co-creation. Essentially, this resulted from visionary thinking and building outside the system.

Given that the system evolves, people's expectations change, and technology changes, our ability to design the future is actually limited. This is especially true if we cannot clearly define the present; the future becomes even more blurred. A vision here is necessary, yet so is the understanding of what can and cannot be influenced; otherwise, the value proposition from the innovation is likely to have a short half-life. The more the future technologies and people are better understood, the more likely the explorations will lead to value propositions with longer half-lives or one solution that can be adapted for future situations. Here actionable foresight helps to ground the innovation that future technology promises to new opportunities. Nevertheless, imagination is needed to make them actionable today to overcome some of the existing challenges.

Actionable foresight is a form of innovation that allows us to anticipate and prepare for future events to take proactive steps to achieve our desired outcomes (Buehring & Liedtka, 2018). Actionable foresight helps organizations identify emerging trends and opportunities they may not have considered otherwise. Foresight links the four horizons, connecting the past with possible futures. To be actionable, foresight must go beyond mere speculation about the future and provide possible concrete scenarios (van der Heijden, 2005). To achieve this, organizations need to have access to a range of data sources, including historical data, market trends, and emerging technologies. They must also have the tools and expertise necessary to analyze this data and identify patterns and trends that can inform their strategic planning. By analyzing data and trends across multiple sectors, organizations can gain new insights into emerging markets and technologies that could help them to achieve their strategic goals.

To attain actionable foresight, organizations must have a culture of continuous learning and innovation (Rohrbeck & Kum, 2018), and at times creative destruction coupled with entrepreneurial discovery (Chiles et al., 2007). This means investing in training and development programs that enable employees to stay updated with the latest trends and technologies and encourage experimentation and risk-taking. To put it bluntly, understanding the present and anticipating the future requires a holistic and continuous approach to innovation that values co-creation, systemic thinking, and actionable foresight.

4 Designing for the Future: The Paradigm Shift from User-Centered to Human and Planet-Centered Design

In the past two decades, the design industry has focused on user-centered design and user experience, moving away from a technology-focused approach (Ritter et al., 2014). This shift has been widely adopted in Design Thinking, Service Design, and UX design, resulting in breakthrough innovations. However, the focus on end-users and oversimplification of their roles has limited the segmentation and insights needed to develop and test solutions effectively.

We must focus on individual actors, roles, and situations to address this limitation, considering the human and planet-centered approach while ensuring profitable, sustainable businesses (Fig. 5). Personas and empathy maps remain core tools for understanding and reframing the problem from multiple perspectives (Stickdorn

Fig. 5 The use of generic user is focused on consumption, and we need to move to become human and plant-centered (artwork by Maryam Afshar of "Studio401")

et al. (2018); Lewrick et al., 2018). However, we must enrich the problem space by including a more comprehensive range of actors and a deeper understanding of their motivations.

While focusing on the benefit for the user is crucial, we must also integrate the planet and broader community into the problem space and solution-building/testing. We can use tools such as lifecycle impact assessment to describe the damage we create over time, but these remain complex to complete. To account for the dynamic nature of the system, we need to consider how the actors evolve over time. A system-wide paradigm shift from user-centered to human- and planet-centered design is needed to address our complex challenges. By focusing on individual actors, roles, and situations, we can create innovative and sustainable solutions that benefit humans, our ecosystem, and the planet.

5 Understanding Value Co-creation in a Connected World

Value co-creation is a process that occurs throughout the lifecycle of a product or service, rather than just during the design phase, and is commonly described as open innovation (Lusch & Nambisan, 2015). Figure 6 illustrates how value co-creation

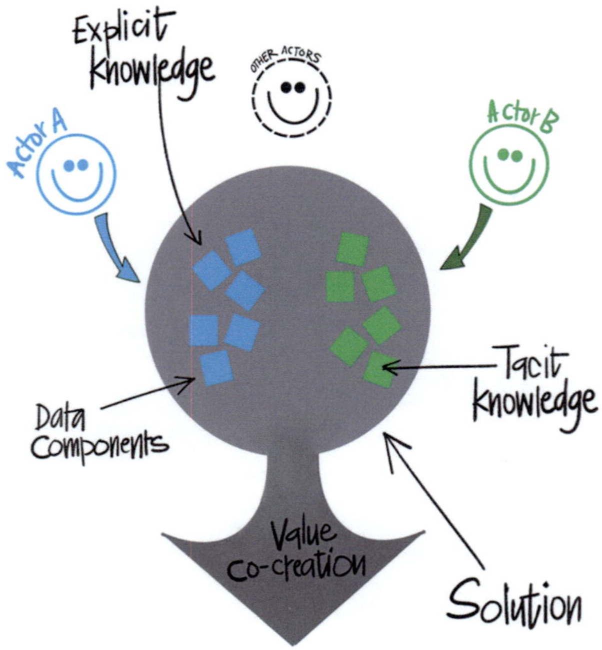

Fig. 6 Value co-creation based on a bundle of knowledge integrated with data (artwork by Maryam Afshar of 'Studio401', based on Valtakoski, 2017)

happens when two or more actors combine resources such as data and knowledge to create a solution together and co-create value for the beneficiary. As we increasingly interact with machines in our connected world, these interactions between actors transform data into information through information design.

This shift from passive users to value co-creation represents a paradigm shift in how we view value. Value is no longer produced and sold to customers but rather co-created with them and other partners (Prahalad & Ramaswamy, 2004). To fully embrace this shift, we must create empathy maps and personas for all actors involved in the innovation process and capture the new knowledge created as they interact. This growth through learning over time is essential in our interconnected world. We can create a more sustainable and value-driven future by assisting customers in their own value-creation processes and viewing them as extended resources.

Empathy goes a long way in creating this co-creation. With a model of innovation based on value creation (e.g., real open innovation), we move to one where we need to understand all the actors in the system. This means creating empathy maps and personas for all actors involved in the innovation process. Also, as they interact together, new knowledge is created, which must be captured within the process. In effect, we grow through learning over time in our interconnected world.

6 Innovating Beyond the Myopic View: Navigating the Interconnected World from Micro to Macro

Here is where individual (micro-level) actions can be supported or hindered by social norms. Zooming out from the individual actors at the micro level, teams and firms compose the meso scale level, and we move to the motivation of firms (Leroy et al., 2013). Here social norms form part of the dynamic system as it evolves over time: we all are influenced by the system, and individuals influence the system.

One approach to overcoming this myopic view has been integrating the "triple bottom line" to rebalance the short-term focus of economics (Elkington, 2013). The meso and macro levels have the most significant impact on the planet as they contain the social and cultural norms that generally evolve slowly with time (Fig. 7). Here is where the economic value focus is contained, and individual economic goals can drive individual motivations. This often results in the maximization of short-term shareholder value that impacts the macrosystem that includes the planet. Similar examples have been seen with dealing with global heating, pensions, and healthcare, all of which can be put off for "one more year," as at the micro and meso levels a single event has no impact at the macro level.

An ambidextrous organization that can balance short-term economic demands with long-term sustainability arguments is needed. Without such a balanced view, the consequences of the "bads" (i.e., "forever chemicals" or global warming) risk new innovations ruining our world (as we know it). Interestingly, in the long run, many of the more ambidextrous firms tend to be more successful and adaptable to

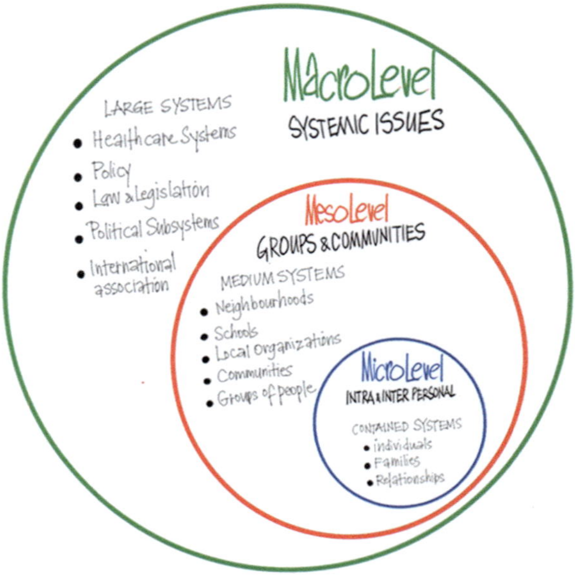

Fig. 7 Learning to zoom in and out between the different layers leads to better innovation (artwork by Maryam Afshar of "Studio401")

the emerging challenges they face as they can embrace them better and let go of the past (O'Reilly & Tushman, 2013). IBM and Toyota are two examples of ambidextrous firms that can reinvent themselves and adapt to the ever-changing world through continued innovation.

Observation is vital to interactive design (McCracken, 1988), yet we need more tools to capture the detailed interactions and motivations that drive them. To overcome the challenge here, we need to learn to observe the world at different levels, from the individual interactions at the micro level up to understanding the problem and testing the solutions at the meso and macro levels. This builds new dimensions into the innovation process and provides a rich understanding of context integrated with the system's dynamics. Too often, innovation falls back here to assumptions, or "for the general good, the user will…", both of which are naïve and lead to poor outcomes.

7 Beyond Design Thinking: Redesigning Design Thinking for a Better Tomorrow

When it comes to improving the innovation process, it is crucial to reflect on how we got here and what the future holds. Despite the initial hype surrounding Design Thinking, its 35-year history and current mainstream acceptance make it a solid foundation to build upon. The evolution of this approach over time has shown its adaptability and effectiveness in driving both disruptive and incremental innovation, showing its merit.

However, as the title suggests, it is time to "Redesign Design Thinking" to continue driving progress and create more impactful innovations that benefit society and the planet. We propose "RE:Designing" it by making five essential changes:

1. **Spend more time to understand the problem**—Einstein famously said that if he had an hour to solve a problem, he would spend 55 min understanding the problem and 5 min solving it. Investing more time in understanding the problem from multiple perspectives and testing possible solutions is crucial for future innovations. In today's complex interconnected world, understanding the links between problems and solutions is essential to creating sustainable solutions. By rebalancing the retrospective analysis with developing and testing new concepts, we can build a solid foundation for innovation.
2. **Consider past context and possible futures**—while past performance does not guarantee future performance, understanding the past provides important contextual information on the problem and existing preconceptions. It is crucial to consider possible futures through scenarios or other tools that help us analyze the environment and our innovations and the possible futures they create.
3. **Move towards human-centric and planet-centric design**—using the term "user" can be limiting, as it suggests a selfish perspective. Instead, we need to adopt a more egalitarian view that considers people as part of the ecosystems in which they exist, with synergies and "bads" to consider as well. This perspective must extend beyond people to include the planet, and we must understand the unintended consequences of our decisions. While technology may be amoral, we, as the decision-makers, must judge whether the outcomes that accrue from actions may be too harmful in the long run.
4. **Understand value co-creation in a connected world**—today, nothing happens in isolation, and the world is becoming increasingly interconnected. We place increasing value on experience, and value co-creation occurs because of these interactions. Open innovation is fundamental to innovation today, and great care must be taken to understand how, when, and where value co-creation occurs during the innovation process. This includes understanding how resources and spontaneity assimilate with the actors and how they collaborate to co-create value for and with the beneficiaries.
5. **View the world at different levels**—today, we live in a system-of-systems, an interwoven world with connections at different levels and with other institutional arrangements. Learning to view our world in detail and from a helicopter perspective is crucial, as this is increasingly necessary for innovation. Ecologists and sociologists have been doing this for many years, and we need to learn from them. The micro, meso, and macro perspectives provide different insights necessary for innovation, either in understanding the challenges or in building and testing solutions. Over time, the system is dynamic, and changes will occur; some fast, some slow.

7.1 Theoretical Contributions

This paper addresses the limitations of current innovation processes and proposes a more integrated approach that combines Design Thinking with other innovative methodologies. The paper advocates for a deeper focus on understanding the problem space while also incorporating a comprehensive understanding of past and future contexts through exploratory scenarios. The need for human-centric and planet-centric design is emphasized, and the importance of considering the broader consequences of proposed solutions is highlighted. Additionally, the paper suggests incorporating service science approaches to understand value co-creation processes at multiple levels better and enhance the existing innovation processes. By integrating these theoretical contributions, the paper aims to support the continued evolution of Design Thinking as a powerful tool for driving impactful innovation.

7.2 Managerial Implications

Leaders and managers must learn to navigate complex interconnected systems of people, processes, and things through sustainable and collaborative leadership and teamwork. They must zoom in and out as needed, understanding the changes and the future impacts that innovations can have. They need to focus on innovative and sustainable behaviors as much as the tools. Innovation will only be actioned as a drive towards a sustainable and collaborative vision. This means that managers need to address four key elements in their future innovation strategies:

1. **More robust networks increase understanding**—in an interconnected world, individuals, teams, and organizations must collaborate to see, know, manage, and address the whole space they need to operate in. We can increase our understanding only through collaboration and the facilitation of increased collaboration and interconnectedness between people, teams, and other organizations for professional and social exchanges.
2. **More co-creation and creativity to increase learning and doing**—building on these new and improved networks, organizations need to leverage their potential to learn and create more through collaboration and creativity. Encouraging experimentation fosters a culture where failure is not viewed as a setback but rather as a necessary learning step. Moving away from the belief that the first attempt should always be perfect and instead embracing a continuous improvement mindset is key. The way forward is to increase volume—more small, low-cost experiments, prototypes, and minimum viable products—to learn and move forward. In a maze, the most critical element is to "move fast and fix things" and stay open to creative and out-of-the-box solutions (Foelsing & Schmitz, 2021).
3. **More outcomes, less output**—prioritize achieving the desired results over producing excessive work. Focus on clear objectives that align with your organization's long-term vision. This shift means less bureaucracy and less work for the

sake of work. Take the time to develop a concrete goal not solely based on monetary gain or market share. Allow projects to develop further and evolve, and if necessary, make the tough decision to shelve them with complete transparency and as needed. Emphasize the potential for positive change and growth instead of focusing on the negative aspects of cutting unproductive projects (Snowden, 2010).

Acknowledgments Many thanks to Maryam Afshar of Studio401, who created the artwork in double quick time based on our rather bad descriptions and prompts.

References

Anderson, J. C., & Narus, J. A. (1998). Business marketing: Understand what customers value. *Harvard Business Review*, pp. 76, 53–67.

Buehring, J. H., & Liedtka, J. (2018). Embracing systematic future thinking at the intersection of strategic planning, foresight and design. *Journal of Innovation Management, 6*(3), 134–152.

Carleton, T. L., & Cockayne, W. R. (2023). *Building moonshots: 50+ ways to turn radical ideas into reality*. John Wiley & Sons.

Chiles, T. H., Bluedorn, A. C., & Gupta, V. K. (2007). Beyond creative destruction and entrepreneurial discovery: A radical Austrian approach to entrepreneurship. *Organization Studies, 28*(4), 467–493.

Dawson, P., & Daniel, L. (2010). Understanding social innovation: A provisional framework. *International Journal of Technology Management, 51*(1), 9–21.

De Bono, E. (2009). Edward de Bono on innovation vs creativity how's it different. *The Business Voice*. Retrieved from https://youtu.be/eWM1PK2XrJ4

Dedehayir, O., & Steinert, M. (2016). The hype cycle model: A review and future directions. *Technological Forecasting and Social Change, 108*, 28–41.

Dodgson, M., Gann, D. M., & Phillips, N. (Eds.). (2013). The Oxford handbook of innovation management. OUP Oxford.

Elkington, J. (2013). Enter the triple bottom line. In *The triple bottom line: Does it all add up?* (pp. 1–16). Routledge.

Foelsing, J., & Schmitz, A. (2021). Check-out. In *New Work braucht New Learning*. Springer Gabler. https://doi.org/10.1007/978-3-658-32758-3_13

Keeley, L., Walters, H., Pikkel, R., & Quinn, B. (2013). *Ten types of innovation: The discipline of building breakthroughs*. John Wiley & Sons.

Leroy, J., Cova, B., & Salle, R. (2013). Zooming in vs zooming out on value co-creation: Consequences for BtoB research. *Industrial Marketing Management, 42*(7), 1102–1111.

Lewrick, M., Link, P., & Leifer, L. (2018). *The design thinking playbook: Mindful digital transformation of teams, products, services, businesses and ecosystems*. John Wiley & Sons.

O'Reilly, C. A., & Tushman, M. L. (2013). Organizational ambidexterity: Past, present and future. *Academy of Management Perspectives, 27*(4), 324–338.

Lusch, R. F., & Nambisan, S. (2015). Service innovation. *MIS Quarterly, 39*(1), 155–176.

McCracken, G. (1988). *The long interview* (Vol. 13). Sage. https://doi.org/10.4135/9781412986229

Pinchot, G., III. (1985). *Intrapreneuring: Why you don't have to leave the corporation to become an entrepreneur*. University of Illinois at Urbana-Champaign's Academy for Entrepreneurial Leadership Historical Research Reference in Entrepreneurship.

Prahalad, C. K., & Ramaswamy, V. (2004). Co-creation experiences: The next practice in value creation. *Journal of Interactive Marketing, 18*(3), 5–14.

Ritter, F. E., Baxter, G. D., Churchill, E. F., Ritter, F. E., Baxter, G. D., & Churchill, E. F. (2014). User-centered systems design: a brief history. In *Foundations for designing user-centered systems: What system designers need to know about people* (pp. 33–54).
Rohrbeck, R., & Kum, M. E. (2018). Corporate foresight and its impact on firm performance: A longitudinal analysis. *Technological Forecasting and Social Change, 129*, 105–116.
Sawaguchi, M. (2011). Innovation activities based on s-curve analysis and patterns of technical evolution–"From the standpoint of engineers, what is innovation?". *Procedia Engineering, 9*, 596–610. https://doi.org/10.1016/j.proeng.2011.03.145
Snowden, D. (2010). *The cynefin framework*. YouTube video, 8, 38. Retrieved from https://www.youtube.com/watch?v=N7oz366X0-8
Stickdorn, M., Hormess, M. E., Lawrence, A., & Schneider, J. (2018). *This is service design doing: Applying service design thinking in the real world*. O'Reilly Media, Inc.
Valtakoski, A. (2017). Explaining servitization failure and deservitization: A knowledge-based perspective. *Industrial Marketing Management, 60*, 138–150. https://doi.org/10.1016/j.indmarman.2016.04.009
Van der Heijden, K. (2005). *Scenarios: The art of strategic conversation*. John Wiley & Sons.
Vargo, S. L., & Lusch, R. F. (2016). Institutions and axioms: An extension and update of service-dominant logic. *Journal of the Academy of Marketing Science, 44*, 5–23.

Shaun West gained a Ph.D. from Imperial College in London and worked for over 25 years in several businesses related to industrial services. He started his industrial career with AEA Technology before moving to National Power, where he developed and sold services to external companies. After an MBA at HEC (Paris), he moved to GE Energy Services, modeling and negotiating long-term service agreements. At Sulzer, he drafted the strategy that led to the service division tripling in size over 10 years and executed part of the strategy by acquiring a 220 M CHF service business. Now at the Lucerne University of Applied Sciences and Arts, he is the Professor of Product-Service System Innovation. He focuses his research on supporting industrial firms to develop and deliver new services and service-friendly business models. He is passionate about exploring complex systems to understand problems better and build solutions faster. He is a member of the advisory board for ASAP Service Management Forum and a member of the Swiss Alliance of Data-Intensive Services. Shaun lives close to Zurich with his wife and two children. He climbs, skis, and runs.

Thomas Schönweitz is a seasoned expert in design thinking and innovation and is a systemic coach. He is the founder and CEO of Whitespring, a boutique consultancy specializing in strategic innovation and organizational development. He has worked with numerous companies, ranging from small startups to multi-billion-euro corporations, helping them navigate the innovation landscape's complexities successfully. Throughout his career, Thomas has demonstrated a keen ability to identify and leverage the potential of human-centered design thinking, driving long-term improvements and building a more sustainable future for all stakeholders involved. With a deep commitment to fostering collaboration and co-creation, Thomas has built up numerous innovation accelerators, corporate innovation labs and programs, empowering individuals and organizations to unlock their full potential and reimagine the future. Thomas is a senior lecturer at HfG Gmünd and a Co-Founder of LifeLongLearning4Future. In his free time, he mentors young founders and start-ups and hosts a podcast on "innovation. better."

Arman Amin, an industrial designer and product designer, has gained over a decade of experience working in various roles with some of the most innovative companies in the world, including Microsoft, Apple, and Tesla. Arman has been sharing his expertise globally as a tutor and mentor, guiding students toward reaching their full potential and achieving their personal and professional goals. In response to the pandemic, Arman established "Send More Brains Inc.," a design firm with an international reach that collaborates with award-winning startups, corporations, and teams to take on complex challenges and drive growth and innovation from within. Arman primarily focuses on supporting his clients in developing strong branding and graphic assets and creating exceptional product and service designs tailored to their specific needs. With an optimistic outlook and futurist mindset, Arman collaborates with innovators worldwide to develop cutting-edge solutions that address some of the world's most pressing issues. A former varsity fencing captain, Arman enjoys longboarding, running, snowboarding, and Yoga and is a major proponent of the educational power of gaming.

Inspired by DARPA: A Global Comparison of Radical Innovation Government Agencies

Tamara Carleton and William R. Cockayne

Abstract The Defense Advanced Research Projects Agency (DARPA) in the United States is renowned for its success in funding radical innovation. This chapter provides an overview of various government agencies that have adopted the core DARPA model outside the defense space in a wide spectrum of areas, such as for homeland security (HSARPA), intelligence (IARPA), energy (ARPA-E), health (ARPA-H), and climate (ARPA-C) in the United States, as well as several other notable examples in Europe and Asia, such as JEDI and ARIA. The chapter compares these groups to the DARPA model and identifies key differences that can offer lessons to others seeking similar outcomes. The chapter also notes the existence of other agencies that were missed due to time constraints, so that a deeper analysis may continue of government agencies around the world pursuing radical innovation.

1 Introduction

The global research and development (R&D) community has long held DARPA as a role model of radical innovation. As one international newspaper reported, "It is the agency that shaped the modern world, and this success has spurred imitators" (The Economist, 2021). Established in 1958 by the US government, DARPA—short for the Defense Advanced Research Projects Agency—has had a mission to find and fund high-risk, high-reward innovation within the US Department of Defense. Originally named as ARPA and later adding the D for its defense focus, DARPA has pursued this vision for over six decades with far more successes than failures, which

T. Carleton (✉)
Department of Mechanical Engineering, Blekinge Institute of Technology, Karlskrona, Sweden
e-mail: tamara.carleton@bth.se

W. R. Cockayne
Institute for Information Management, University of St. Gallen, St. Gallen, Switzerland
e-mail: william.cockayne@unisg.ch

T. Carleton et al. (eds.), *Inventing the Almost Impossible*, Future of Business and Finance, https://doi.org/10.1007/978-3-031-36224-8_3

is highly unusual for any organization and exceptional for those operating at the fuzzy front end of innovation. Although DARPA's primary focus has been on new technologies and solutions for the American military forces, various inventions that the agency has envisioned and funded—such as the internet precursor called ARPANET (Hafner & Lyon, 1998; Waldrop, 2018), the Global Positioning System (GPS), and mRNA vaccine technology that led to the COVID-19 vaccine (Sonne, 2020)—have been adopted by many groups outside the defense sphere, leading to DARPA's outsized impact in the world spanning multiple areas from global communication and transportation to medicine and trade. As such, multiple experts from inside the agency (e.g., Dugan & Gabriel, 2013) and outside the agency (e.g., Bonvillian, 2019) have presented their view of the special DARPA formula and how its unique model can apply to contexts far outside defense applications. Moreover, other groups within and outside government have been inspired by DARPA to set up similar efforts for funding breakthrough research into development.

This chapter builds on prior analysis about DARPA and examines a spectrum of other radical innovation groups—many of whom even embed ARPA in their organizational name—that have adopted the DARPA model in some fashion. The spectrum considers more established groups, such as ARPA-E (energy), to more recent groups, such as ARPA-H (health) and ARIA (United Kingdom), focusing predominantly in the public sector. How do all these groups compare to the classic DARPA model, and what have they done differently that can offer lessons to others seeking similar outcomes? We will first consider DARPA and then describe the other groups in turn. In addition, we have opted not to call these DARPA-inspired agencies as *clones*, as some of them have changed key elements in the classic DARPA model. Instead we call them *cousins* because they share family blood so to speak, yet live in a different part of the overall family tree. Table 1 summarizes the main groups discussed in the chapter.

Table 1 A summary of DARPA and its organizational cousins in the US and non-US public sector

Organization	Established	Nation	Focus
DARPA	1958	USA	Defense
DARPA Cousins in the US Public Sector			
HSARPA	2002	USA	Homeland security
IARPA	2007	USA	Intelligence
ARPA-E	2009	USA	Energy
ARPA-H	2022	USA	Health
ARPA-C	Proposed	USA	Climate
DARPA Cousins in the non-US Public Sector			
Tekes	1983	Finland	Technology
JEDI	2018	Europe	Technology
Moonshot R&D	2019	Japan	Technology
SPRIN-D	2019	Germany	Technology
Cybersecurity Innovation Agency	2020	Germany	Defense
ARIA	2022	UK	Technology
CARPA	Proposed	Canada	Technology

2 About the Defense Advanced Research Projects Agency (DARPA)

Established in 1958, DARPA is an independent federal agency situated under the US Department of Defense (DOD). Originally named as the Advanced Research Projects Agency (ARPA), it was soon renamed as DARPA due to its defense mandate. The agency's goal is to fund breakthrough technologies for national security, or as one *Science* writer put it, "DARPA's job is to make sure the US military holds a technological edge over its enemies" (Mervis, 2016, para. 4). DARPA has approximately 220 government employees that span six technical offices, including nearly 100 program managers, who together oversee about 250 R&D programs at a given time (DARPA, 2023). These program managers (PMs) are instrumental in deciding which technology areas to pursue and in recruiting new PM talent with pioneering tech visions (Carleton, 2010). Several key structural elements in the agency enable the autonomy of these PMs, namely a project-based structure (hence the P in the DARPA name), lightweight governance (especially no formal peer review or committee decision making for program approvals), and regular turnover (PMs are typically hired as government employees in nonrenewable contracts averaging 4 years in length, unlike most career employees working at sister organizations such as the National Science Foundation or National Institutes of Health in the United States). See a comprehensive discussion of DARPA in the book entitled *The DARPA Model for Transformative Technologies*, available open source (Bonvillian et al., 2019).

DARPA's budget was US $3.8 billion in fiscal year 2022. At least four primary factors contribute to DARPA's ability to transfer its knowledge to the next innovation stage—crucial in bridging the gap between basic research and its application. First, DARPA relies heavily on procurement programs from the American military services, which offer an early testbed and user base for DARPA ideas. Subsequently, DARPA PMs do not spend as much time finding and convincing early adopters as typically expected in other types of organizations. Second, there is a strong expectation culturally within DARPA that the teams it funds—referred to as "performers" within DARPA—produce prototypes and proofs of concept, not publications or patents. As *The Economist* notes, "Whereas most [organizations] focus on basic research, DARPA builds things" (The Economist, 2021, para. 5). DARPA PMs frequently interact with the performers that they fund, providing constant feedback that help to clarify both the project idea and related metrics of success throughout a project. This interaction differs from traditional funding models, which are often treated as "won and done" (Piore et al., 2019). Third, DARPA does not own its own labs or intellectual property, instead enabling the teams that it funds to decide accordingly. This setup creates a strong incentive for DARPA-funded teams to advance new ideas on their own financial and professional terms. Fourth, beyond creating and demonstrating potential new technologies, DARPA also creates new technical communities and new academic fields as a result of discussing new ideas—fields such as materials science, computer science, and autonomous systems (Piore et al., 2019; Van Atta et al., 2019). Participants in these communities frequently continue work started by DARPA through their teaching, additional R&D, and even commercialization, all of which further contributes to radically innovative ideas moving forward in the world. This combination has been at DARPA's organizational core across decades and offers a loose checklist for other groups seeking to emulate DARPA's success.

Although some critics (e.g., Gross, 2018; Jacobsen, 2015) have worried about the agency being able to balance pushing the boundaries of technology while considering ethical and moral implications, DARPA has consistently produced major technological breakthroughs for over 65 years. In fact, one study by researchers at the Massachusetts Institute of Technology (MIT) found that DARPA performed better in terms of patents, tech licenses, and new business ventures compared to multiple US agencies, including the National Science Foundation, despite the fact that DARPA explicitly rejects these performance measures (Piore et al., 2019). Other scholars (e.g., Azoulay et al., 2019) have discussed that other ARPA-like agencies will be similarly challenged when measuring long-term transformation.

3 DARPA Cousins in the US Public Sector

We consider five agencies in the American government deliberately established with a mission of radical innovation between 1958 and 2022. As we sought agencies modeled after DARPA or have been actively compared with DARPA in the news, we may have overlooked agencies in other parts of the US public sector that share a similar mission yet keep a low profile. We discuss the agencies HSARPA, IARPA, ARPA-E, ARPA-H, and ARPA-C in order of official operational date.

3.1 About the Homeland Security Advanced Research Projects Agency (HSARPA)

Authorized by the US Homeland Security Act of 2002, the Homeland Security Advanced Research Projects Agency (HSARPA) was established in the US Department of Homeland Security (DHS). HSARPA's mission was to support "cutting-edge research to produce revolutionary changes in technologies, new capabilities and threat and risk assessments," primarily in DHS's division of Homeland Security Enterprise (US Department of Homeland Security, 2023). Initial hiring positions were limited to 20 total. HSARPA's budget started with roughly US $1 billion, which supported programs across five main divisions. Given the name association with DARPA and bold language in its stated mission, multiple experts expected HSARPA to pursue similar breakthrough efforts. Instead, DHS leadership reduced HSARPA's focus to conduct "essentially all of its [DHS] extramural activities, most of which were conventional R&D with only moderate risk" (Shea &

Morgan, 2009, p. 14). Its early staff—many of whom came from DARPA, although not the acting director or deputy director—left in frustration because HSARPA was not adequately supported nor allowed to operate as a genuine separate unit within the DHS (Bonvillian, 2019; Nieto-Gómez, 2011). By 2009, DHS's Undersecretary for Science and Technology tried to revive HSARPA back to its original bold mission with limited success (Bonvillian, 2019). As of 2023, the HSARPA webpage on the DHS website is marked as archived content (US Department of Homeland Security, 2023).

3.2 About the Intelligence Advanced Research Projects Agency (IARPA)

As the agency name suggests, IARPA is modeled after DARPA. Established in 2007, the Intelligence Advanced Research Projects Agency is tasked with funding cutting-edge technologies in American national intelligence. Its first director was a former DARPA PM who joined in 2008, who deliberately sought to replicate much of DARPA's ways of working (Bonvillian, 2019). By 2019, IARPA had 25 program managers. Most of its research work is unclassified with no publication restrictions. However, IARPA's budget, like most of the overall US intelligence budget, remains classified information (American Psychological Association, 2022). IARPA has a similar high record of success to DARPA; one agency director stated that 70% of IARPA programs have achieved at least one tech transition to an intelligence agency (Bonvillian, 2019).

At least three major differences exist between DARPA and IARPA (ibid.). First, IARPA faces longer hiring processes for its PMs due to the need for high-level security clearances, which often take several months or longer. Second, IARPA organizes its programs as *tournaments*, in which multiple teams are funded in parallel following a common set of metrics. IARPA then spends approximately a quarter of its budget on independent testing and evaluation, which includes an internal position for a Chief of Testing and Evaluation. This testing process results in at least one team cut per program phase, sometimes even discontinued programs. Third, IARPA lacks a large user base in the intelligence community like DARPA has across the DOD. As a result, IARPA has taken extra care to transition new technologies into various US intelligence agencies, an effort overseen by a full-time Chief of Technology Transition. In addition, an IARPA program typically develops a tech transition plan during its second or third year, which helps boost tech transfer.

3.3 About the Advanced Research Projects Agency-Energy (ARPA-E)

Inspired directly by DARPA's model, ARPA-E has a similar mission to fund high-risk, high-reward ideas in the energy space. Formally named as the Advanced Research Projects Agency–Energy, ARPA-E is a government agency under the US Department of Energy (DOE) tasked with funding advanced energy technologies and bridging the gap between basic research and commercial development. Though officially created in 2007 by the America COMPETES Act, ARPA-E was established 2 years later with an operating budget of US $400 million as part of the 2009 American Recovery and Reinvestment Act (ARPA-E, 2023a, "Budget"). Its early staffing included two DARPA veterans among its eight PMs, who helped to transfer institutional knowledge about DARPA to the new energy entity (Bonvillian & Van Atta, 2019). The number of PMs then grew to 15 total over the next 10 years. Like DARPA, ARPA-E can sidestep the usual cumbersome US governmental hiring process using a special civil service waiver that was included in the America COMPETES Act (ibid.). However, by its first decade of operations, ARPA-E has functioned more as a single DARPA office in terms of size and has not yet added a second layer of office directors. ARPA-E has taken pains to promote its impact beyond projects, which many feel that DARPA can ignore more easily given its agency tenure and track record. For example, ARPA-E published "impact sheets" in 2016, 2017, and 2018 that summarize agency outcomes, and its first impact sheet sampled project outcomes since its founding (ARPA-E, 2023b, "Our").

Bonvillian and Van Atta (2019) describe several major differences between DARPA and ARPA-E. ARPA-E program managers are limited to 3-year contracts, compared to DARPA's contracts which range between 3 to 5 years, thus creating a tighter timeline for action. Unlike DARPA, ARPA-E has introduced a small Fellows program of recent PhD graduates who support the PMs with additional technical expertise. In addition, ARPA-E uses a two-stage selection process to approve new projects, which creates a feedback cycle in the first stage for applicants to respond and clarify key points. This process has allowed for more collegial discussions to occur early in the decision-making process between ARPA-E PMs and their performers. ARPA-E PMs also tend to rely more on expert panels to guide their project decisions than DARPA has historically done.

In terms of tech handoff, ARPA-E lacks the same type of captive internal market within the DOE, which DARPA finds in the DOD. ARPA-E also faces hefty competition against 17 national energy laboratories and 46 Energy Frontier Research Centers, among other established groups across the American energy landscape (Bonvillian & Van Atta, 2019). To address this gap, ARPA-E has engaged key energy stakeholders to serve on its review panels and introduced a "tech-to-market" program with follow-on seed grants designed to help their funded teams develop additional commercial promise. By the late 2010s, ARPA-E had explored how to expand their user base, including an early partnership with the DOD, as military groups often have pressing needs related to energy.

3.4 About ARPA-Health (ARPA-H)

In 2017, two former US federal researchers proposed the concept of HARPA as a health-oriented agency modeled after DARPA, which then influenced growing discussions within the broader community of American health scholars and policymakers (Suzanne Wright Foundation, 2021). By 2022, the US government established ARPA-Health, dubbed as ARPA-H, as an independent entity within the US National Institutes of Health (NIH) that funds pivotal investments to "tackle the hardest challenges in health" (ARPA-H, 2023a) and "benefit the health of all Americans by catalyzing health breakthroughs that cannot readily be accomplished through traditional research or commercial activity" (The White House, 2021b). Both ARPA-H's agency director and acting deputy director have direct prior knowledge serving as former PMs in IARPA and DARPA, allowing them to draw from their personal experiences (NIH Record, 2022a, 2022b). A deputy director joining ARPA-H in 2023 also brings prior experience from HSARPA (ARPA-H, 2023b). ARPA-H launched with a US $1 billion budget, which was increased to US $1.5 billion for fiscal year 2023 (ARPA-H, 2023c). As ARPA-H is still in its infancy as an agency, there is little published information about its internal organization or projects at the time of writing. That said, various experts at NIH and other organizations (e.g., Collins et al., 2021; Thorp, 2022) have offered their recommendations on setting ARPA-H up for success.

3.5 About the Advanced Research Projects Agency-Climate (ARPA-C)

The concept of a DARPA-inspired entity for climate innovation has existed for some time. For example, David and Lucile Packard Foundation gave a grant to a California-based environmental consultancy in 2015 to explore how philanthropies might launch their own versions of a climate-focused DARPA (Iaconangelo, 2020). This work soon led to the Climate Breakthrough Project, in which five foundations collectively fund "large, multi-year, unrestricted awards to help empower promising leaders with powerful, high-risk, high-reward innovations in the climate space" (Climate Breakthrough Project, 2023). Separately in 2019, several professors from Colorado State University proposed the concept of ARPA-C to US presidential candidate Joe Biden as a radical model to address intractable climate problems (Badia et al., 2021). Their white paper informed federal discussions, and by 2021, the Biden administration announced a new Climate Innovation Working Group as a step toward launching an Advanced Research Projects Agency-Climate (ARPA-C) in the US Department of Energy (The White House, 2021a). As part of the American Jobs Plan announced in 2021, ARPA-C's mission is "to develop new methods for reducing emissions and building climate resilience, as well as expanding across-the-board funding for climate research" (Department of Energy, 2021). ARPA-C has a total proposed initial budget of US $500 million, and the American Jobs Plan proposes an additional one-time infusion of US $15 billion (Peterson, 2021). ARPA-C would mirror ARPA-E's innovation model more so than DARPA (ibid.). While some policymakers and scientists have been concerned of the potential overlap between ARPA-E and ARPA-C, ARPA-C aims to focus on climate-related innovation in areas other than energy. However, there has been scant public coverage on this agency proposal since 2021, and a 2023 keyword search on the DOE website returns no updated results for ARPA-C.

4 DARPA Cousins in the Non-US Public Sector

Outside of the United States, several agencies have been inspired by the DARPA model for radical innovation. We discuss several notable examples from Europe and Asia in order of founding date, which comprise Tekes in Finland, JEDI in France/Europe, the Moonshot R&D Program in Japan, SPRIN-D in Germany, the Agency for Innovation in Cybersecurity in Germany, ARIA in the United Kingdom, and the CARPA concept in Canada. We did not include other DARPA aspirants, such as France's Defense Innovation Agency set up in 2018, due to time constraints.

4.1 About Tekes (The Finnish Funding Agency for Technology and Innovation) in Finland

In Finland, Tekes (The Finnish Funding Agency for Technology and Innovation) was founded in 1983, directly inspired by the DARPA model to fund areas of new growth within Finland (personal communication, Raine Hermans, 2010). It began with 20 people, and part of its mission was to foster public-private R&D networks. Situated under the Finnish Ministry of Employment and the Economy, Tekes was soon seen as the most important public funding agency for research funding in Finland. In 2009, Tekes was actively funding nearly 2200 projects with a total budget of €579 million (Helsinki Partners, 2011). At this point, the agency employed over 380 people and had six offices abroad in Beijing, Brussels, Tokyo, Shanghai, Silicon Valley, and Washington DC (ibid.). By 2018, the Finnish government merged Tekes with another government agency called Finpro, a trade promotion agency, to create a new entity called Business Finland. Since then, Business Finland has been focused more on attracting foreign talent and investment rather than funding technological breakthroughs (personal communication, Christopher Palmberg, 2023).

A sister organization in Finland worth mentioning is Sitra, the Finnish Innovation Fund, which is an independent public foundation that operates under the supervision of the Finnish Parliament. Founded in 1967 originally as a part of the Bank of Finland on the country's 50th anniversary, Sitra began with an endowment capital of

FIM 100 million (Karo & Kattel, 2016). Although Sitra's mission focuses more on near-term investments for business creation than technological breakthroughs, the foundation shares some similar organizational features with DARPA. For example, in 2012, Sitra shifted from a program structure to a looser matrix model based on projects and themes. Sitra employs approximately 100–120 workers with nearly half on fixed-term contracts. In addition, the foundation has similar autonomy to DARPA for its program and funding decisions; however, its funding structure differs. Unlike DARPA, Sitra does not depend on federal funds and instead derives its budget from its endowment, which averages a return of approximately €30–40 million annually (Sitra, 2023).

4.2 About the Joint European Disruptive Initiative (JEDI) in Europe

The Joint European Disruptive Initiative (JEDI) bills itself as "The European ARPA." It was set up by the French-based founder of an investment fund in 2018 to accelerate the leadership of France and Germany in disruptive innovations (Detroy, 2019). JEDI started with grand challenges, which were mostly financed by foundations, philanthropists, and donors across France and Germany, with the intention that by 2021, its projects would become European programs (Mawad, 2020). Technically structured as a nonprofit, JEDI has actively sought pan-European governmental support to operate at a similar scale as DARPA. The group's culture mixes the DARPA principles with the spirit of the heroic Jedi characters from the *Star Wars* franchise, including inviting supporters to join in their "JEDI Pledge." Public details about the group's budget and staffing are limited, although JEDI promotes that it is powered by 4600 leaders of Europe's deeptech ecosystem in 29 countries (JEDI, 2023).

4.3 About the Moonshot R&D Program in Japan

In 2019, the Japanese government announced the creation of a new Moonshot R&D program within its Cabinet Office with a budget of US $897 million (¥100 billion) (Mallapaty, 2019). The program aims to "create disruptive innovations from Japan and promotes challenging R&D based on revolutionary concepts that are not simply the extension of existing technologies, i.e. moonshots" (Cabinet Office, 2023). The program began with seven lofty objectives, ranging from freeing people "from limitations of body, brain, space, and time by 2050" to the realization of "sustainable care systems to overcome major diseases by 2040" (Cabinet Public Affairs Office, 2020). All goals are set by the Council for Science, Technology, and Innovation in Japan's Cabinet Office and the Headquarters for Healthcare Policy in the Prime

Minister's Office of Japan with input from external experts. Like DARPA, Japan's Moonshot R&D Program will rely on PMs. Little information or press coverage has been provided outside Japan about the program's progress since its launch.

4.4 About the Federal Agency for Disruptive Innovation (SPRIN-D) in Germany

In 2019, Germany announced the Federal Agency for Disruptive Innovation to fund radical ideas with a budget of around US $1.2 billion (€1 billion) over 10 years. The agency is known by the acronym SPRIN-D as an allusion to a sprint in a race (Kumagai, 2021). SPRIN-D is led by an entrepreneur, considered a pioneer of open-source technology (ibid.). However, press coverage of 2 years (e.g., Rinke & Nienaber, 2021) revealed that the new agency with a staff of 36 people was struggling in the German innovation environment. Unlike DARPA, SPRIN-D faces multiple constraints that limit its staff autonomy. For example, the European Union requires joint rules for state aid and procurement, which complicates funding for any SPRIN-D project (ibid.). In addition, the German Federal Court of Auditors soon ruled that SPRIN-D must follow the same public spending rules as any other semi-state institution (Burke, 2021).

4.5 About the Cybersecurity Innovation Agency in Germany

Soon after SPRIN-D, the German government launched the Agency for Innovation in Cybersecurity in 2020 based on DARPA's model. Like DARPA, this agency focuses directly on national defense, particularly cyberattacks. The new agency started with a budget of US $412 million (€350 million) through 2023 (Deutsche Welle, 2020). Led by an expert in artificial intelligence, the agency planned to hire 100 people. Little public information has been found on this agency since 2021.

4.6 About the Advanced Research and Invention Agency (ARIA) in the United Kingdom

In 2019, Dominic Cummings, then the prime minister's chief adviser, provoked the idea of a DARPA-like agency for the United Kingdom that could make pivotal investments for inventing the future of Britain (The Economist, 2020). The concept had a controversial reception, but eventually plans were outlined by early 2021. ARIA was ultimately approved in the UK ARIA Act in early 2022 with a budget of US $920 million (£800 million) over 4 years (Gov.uk, 2021, 2022). By mid-2022, the British government appointed an American nonprofit leader as the head of ARIA (Gov.uk, 2022). By late 2023, ARIA hired 8 program directors modeled loosely after DARPA PMs, who have the "freedom to identify and fund transformational science and technology at speed with minimal bureaucracy" (ibid., para. 1). As the British agency noted, "To work out what qualities made an effective programme director, we interviewed around 60 people who worked at or with DARPA and its related agencies, alongside others across the R&D ecosystem. In the UK, we don't have the same cultural precedents or existing networks of ready-to-go programme directors as in the US" (Advanced Research and Invention Agency, 2023, para. 5). While ARIA sits formally within the UK Department for Business, Energy, and Industrial Strategy, the agency is expected to remain independent from UK Research and Innovation, which is its main research funding agency (O'Grady, 2021).

4.7 About the Canada Advanced Research Projects Agency (CARPA) in Canada

In early 2021, policy analyst Robert Asselin at the Business Council of Canada wrote a memo to federal decision makers entitled "Canada Needs an R&D Powerhouse Modelled on the Successes of DARPA" (Asselin, 2021), which then became a lengthy report calling for a "new north star" (Asselin & Speer, 2021). As part of other discussions underway that year, the Liberal Party of Canada proposed a Canada Advanced Research Projects Agency (CARPA) as part of its reelection platform, which was to be modeled after DARPA with an initial budget of CAD $2 billion (Liberal Party of Canada, 2023). Led by Canada's 23rd Prime Minister Justin Trudeau, the CARPA concept has received some media pushback as being unrealistic and misguided (e.g., Wells, 2021). A similar concept in 2021 was floated by the opposition Conservative party, which also proposed setting up a dedicated government agency to fund research into next-generation technologies (Synder, 2021). By spring 2022, the Canadian government appears to have abandoned the idea of CARPA.

5 Conclusion and Further Research

In summary, this chapter provides an overview of various agencies that have adapted the DARPA innovation model across the US and global public sector. Our comparison has been limited largely to secondary data sources and publicly available information in English, so further research should investigate this topic more deeply, especially delving into each agency's internal sources and publications in respective native languages, in order to develop a more complete and comprehensive analysis. One future research angle should analyze the timing for these agencies of radical innovation; for example, why have several DARPA-like agencies launched in the last two decades? Do other government agencies and organizations that share DARPA's mission exist in earlier decades and in other countries not included in this data sample, and how do they compare to this data set? Another research angle could explore how these agencies operate as change agents in their respective innovation ecosystem, as these agencies deliberately seek to provoke paradigm shifts and initiate new technical communities, which can challenge the status quo of existing system actors. Furthermore, other government entities following the DARPA spirit have been proposed in other areas, including agriculture (called AgARPA), the environment (EnARPA), and transportation/infrastructure (TARPA) (e.g., Stebbins & Ling, 2020), all of which offer different possibilities for the future and more comparison points in the DARPA galaxy. A third research angle could explore how DARPA's model has been adapted in industry, particularly by former DARPA PMs setting up radical innovation units in corporations--such as within Intel, Motorola, and Alphabet. By deepening the analysis of DARPA cousins and other organizational variants in different areas, new groups seeking similar outcomes in radical innovation can better define their mission, structure, and results.

References

Advanced Research and Invention Agency. (2023, August 29). *The search for our PDs is over.* Email newsletter.

American Psychological Association. (2022, November 16). *ARPA and you: Research process and funding opportunities at Intelligence Advanced Research Projects Activity (IARPA).* Essential Science Conversations series. https://www.apa.org/science/programs/essential-conversations/research-process-funding-transcript.pdf

ARPA-E. (2023a). *Budget requests.* https://arpa-e.energy.gov/about/budget-requests

ARPA-E. (2023b). *Our impact.* https://arpa-e.energy.gov/about/our-impact

ARPA-H. (2023a). *About us.* https://arpa-h.gov/about/about-arpa-h/

ARPA-H. (2023b, January 6). *ARPA-H selects Dr. Susan Monarez as deputy director* [Press release]. https://arpa-h.gov/news/pr-monarez-deputy-director/

ARPA-H. (2023c). *Budget and appropriations.* https://arpa-h.gov/about/budget/

Asselin, R. (2021). *Canada needs an R&D powerhouse modelled on the successes of DARPA.* C.D. Howe Institute. https://www.cdhowe.org/intelligence-memos/robert-asselin-%E2%80%93-canada-needs-rd-powerhouse-modelled-successes-darpa

Asselin, R., & Speer, S. (2021, December). *New north star III: The case for a Canada Advanced Research Projects Agency.* Public Policy Forum. https://ppforum.ca/wp-content/uploads/2021/12/NNS3-CanadaAdvancedResearchProjectsAgency-PPF-Dec2021-EN.pdf

Azoulay, P., Fuchs, E., Goldstein, A. P., & Kearney, M. (2019). Funding breakthrough research: Promises and challenges of the "ARPA Model". *Innovation Policy and the Economy, 19*(12019), 69–96.

Badia, L., Plaut, J. M., von Fisher, J. C., Volckens, J., & Muhs, J. (2021). Envisioning ARPA-C: A transdisciplinary institution for radical climate research and intervention. *Earth's Future, 9*(6), e2021EF002115. https://doi.org/10.1029/2021EF002115

Bonvillian, W. B. (2019). IARPA: Modified DARPA innovation model. In W. B. Bonvillian, R. Van Atta, & P. Windham (Eds.), *The DARPA model for transformative technologies: Perspectives on the US Defense Advanced Research Projects Agency.* Open Book Publishers.

Bonvillian, W. B., & Van Atta, R. (2019). ARPA-E and DARPA: Applying the DARPA model to energy innovation. In W. B. Bonvillian, R. Van Atta, & P. Windham (Eds.), *The DARPA model for transformative technologies: Perspectives on the US Defense Advanced Research Projects Agency*. Open Book Publishers.

Bonvillian, W. B., Van Atta, R., & Windham, P. (Eds.). (2019). *The DARPA model for transformative technologies: Perspectives on the US Defense Advanced Research Projects Agency*. Open Book Publishers.

Burke, F. (2021, July 15). *German researchers draw up demands for incoming science minister*. Science|Business. https://sciencebusiness.net/news/german-researchers-draw-demands-incoming-science-minister

Cabinet Office, Government of Japan. (2023). *About Moonshot Research and Development Program*. https://www8.cao.go.jp/cstp/english/moonshot/system_en.html

Cabinet Public Affairs Office, Government of Japan. (2020, November 20). Japan's Moonshot Research Program is taking on the biggest challenges. *Forbes*.

Carleton, T. (2010). *The value of vision in radical technological innovation* [Doctoral dissertation, Stanford University]. https://searchworks.stanford.edu/view/8729358

Climate Breakthrough Project. (2023). *Launching bold climate solutions for the planet*. https://www.climatebreakthrough.org/about/

Collins, F. S., Schwetz, T. A., Tabak, L. A., & Lander, E. S. (2021). ARPA-H: Accelerating biomedical breakthroughs. *Science, 373*(6551), 165–167.

Department of Energy. (2021, April 20). *How the American Jobs Plan will advance America's energy and competitiveness priorities*.

Detroy, P. F. (2019, February 18). André Loesekrug-Pietri (JEDI): Le meilleur moyen de prouver que ça marche, c'est d'expérimenter. *Décideurs Magazine*.

Deutsche Welle. (2020, November 8). *Germany launches cybersecurity agency*. https://www.dw.com/en/germany-launches-cybersecurity-agency-to-strengthen-digital-sovereignty/a-54529134

Dugan, R. E., & Gabriel, K. J. (2013). 'Special forces' innovation: How DARPA attacks problems. *Harvard Business Review, 91*(10), 74–84.

Gov.uk. (2021, March 19). *Advanced Research and Invention Agency (ARIA): Policy statement* [Policy paper]. https://www.gov.uk/government/publications/advanced-research-and-invention-agency-aria-statement-of-policy-intent/advanced-research-and-invention-agency-aria-policy-statement

Gov.uk. (2022, July 19). *Innovation heavyweights appointed to lead new Advanced Research and Invention Agency* [Press release].

Gross, M. J. (2018). The Pentagon's push to program soldiers' brains. *The Atlantic*.

Hafner, K., & Lyon, M. (1998). *Where wizards stay up late: The origins of the Internet*. Simon and Schuster.

Helsinki Partners. (2011). *Tekes, the Finnish Funding Agency for Technology and Innovation*. https://helsinkibusinesshub.fi/tekes-the-finnish-funding-agency-for-technology-and-innovation/

Iaconangelo, D. (2020, May 28). ARPA-C? Biden's 100% clean energy plan questioned. *Energywire*. https://www.eenews.net/articles/arpa-c-bidens-100-clean-energy-plan-questioned/

Jacobsen, A. (2015). *The Pentagon's Brain: An uncensored history of DARPA*. America's Top-Secret Military Research Agency.

JEDI. (2023). *The European ARPA*. https://www.jedi.foundation/about

Karo, E., & Kattel, R. (2016). How to organize for innovation: Entrepreneurial state and organizational variety. *Working Papers in Technology Governance and Economic Dynamics*, no. 66.

Kumagai, T. (2021). *Why did the German government create an agency for disruptive innovations?* DWIH Tokyo (German Centre for Research and Innovation Tokyo). https://www.dwih-tokyo.org/en/2021/05/25/sprin-d/

Liberal Party of Canada. (2023). *A new advanced research agency*. https://liberal.ca/our-platform/a-new-advanced-research-agency/

Mallapaty, S. (2019, April 9). Japan prepares 'moonshot' project to solve global problems. *Nature*. https://www.nature.com/articles/d41586-019-01094-w

Mawad, M. (2020, June 5). *Meet the JEDI fighting Covid...and for Europe's tech future*. Sifted. https://sifted.eu/articles/jedi-innovators
Mervis, J. (2016). What makes DARPA tick? *Science, 351*(6273), 549–553.
Nieto-Gómez, R. (2011). The power of 'the few': A key strategic challenge for the permanently disrupted high-tech homeland security environment. *Homeland Security Affairs, 7*(18) https://www.hsaj.org/articles/50
NIH Record. (2022a, June 10). Russell named acting deputy director for new advanced research entity. LXXIV(12).
NIH Record. (2022b, October 14). Wegrzyn named first ARPA-H director. LXXIV(21).
O'Grady, C. (2021, February 26). New U.K. funding agency aims to tackle innovative research. *Science, 371*(6532), 875.
Peterson, A. (2021, June 30). *FY22 budget request: DOE applied energy R&D*. American Institute of Physics. https://www.aip.org/fyi/2021/fy22-budget-request-doe-applied-energy-rd
Piore, M. J., Colatat, P., & Beck Reynolds, E. (2019). NSF and DARPA as models for research funding: A institutional analysis. In W. B. Bonvillian, R. Van Atta, & P. Windham (Eds.), *The DARPA model for transformative technologies: Perspectives on the US Defense Advanced Research Projects Agency*. Open Book Publishers.
Rinke, A., & Nienaber, M. (2021, August 20). Analysis: Red tape, risk aversion clip wings of Merkel's innovation legacy. *Reuters*.
Shea, D. A., & Morgan, D. (2009, June 22). *The DHS Directorate of Science and Technology: Key issues for Congress*. Congressional Research Service, Rep. No. 7-5700: RL34356. https://sgp.fas.org/crs/homesec/RL34356.pdf
Sitra. (2023). *Facts about Sitra*. https://www.sitra.fi/en/topics/facts-about-sitra/
Sonne, P. (2020, July 30). How a secretive Pentagon agency seeded the ground for a rapid coronavirus cure. *Washington Post*.
Stebbins, M., & Ling, G. (2020, April 19). *Creating the Health Advanced Research Projects Agency (HARPA)*. Day One Project. https://www.dayoneproject.org/ideas/creating-the-health-advanced-research-projects-agency-harpa
Suzanne Wright Foundation. (2021). *Our report on HARPA*. https://www.harpa.org/news
Synder, J. (2021, August 11). Conservatives unveil innovation policy ahead of potential election call, marking departure from Liberals. *National Post*. https://nationalpost.com/news/politics/conservatives-unveil-innovation-policy-ahead-of-potential-election-call-marking-departure-from-liberals
The Economist. (2020, January 23). Recreating ARPA, the most successful research agency in history.
The Economist. (2021, June 3). A growing number of governments hope to clone America's DARPA.
The White House. (2021a, February 11). *Biden-Harris Administration launches American innovation effort to create jobs and tackle the climate crisis* [Press release].
The White House. (2021b, September). *ARPA-H frequently asked questions*. https://www.whitehouse.gov/wp-content/uploads/2021/09/092921-ARPA-H-FAQ.pdf
Thorp, H. H. (2022). Will ARPA-H work? *Science, 376*(6590), 223.
US Department of Homeland Security. (2023). *Homeland Security Advanced Research Projects Agency*. https://www.dhs.gov/science-and-technology/hsarpa
Van Atta, R., Windham, P., & Bonvillian, W. B. (2019). Lessons from DARPA's experience. In W. B. Bonvillian, R. Van Atta, & P. Windham (Eds.), *The DARPA model for transformative technologies: Perspectives on the US Defense Advanced Research Projects Agency*. Open Book Publishers.
Waldrop, M. M. (2018). *The dream machine*. Stripe Press.
Wells, P. (2021, September 1). The Liberal platform, or CARPA diem. *Maclean's Magazine*. https://macleans.ca/politics/ottawa/federal-election-2021-the-liberal-platform-or-carpa-diem/

Tamara Carleton is a globally recognized expert in radical innovation. As founder and CEO of Innovation Leadership Group, Tamara works with business and government leadership teams to help them build innovation capability and lay out bold futures. With a penchant for creating new methods and tools, Tamara is also a renowned educator, teaching strategic foresight and innovation at multiple universities worldwide. She holds a doctorate in mechanical engineering from Stanford University, where she explored DARPA's enduring innovation practices. She is the author of *Building Moonshots: 50+ Ways to Turn Radical Ideas into Reality* and the *Playbook for Strategic Foresight and Innovation*.

William R. Cockayne has spent his life building great teams who can imagine, invent, and deliver the future. He is a visionary technologist with a passion for understanding the latest technologies. He holds a doctorate in mechanical engineering from Stanford University and is the co-author of *Building Moonshots: 50+ Ways to Turn Radical Ideas into Reality* and the *Playbook for Strategic Foresight and Innovation*.

Inventing the Future of Aviation

Mina Bastawros and Oliver Haas

Abstract Two highly qualified disruptors in the aviation industry explore the needs, opportunities, and imagined future of air transportation. Combining experience from Airbus and Cathay Pacific, along with intrapreneurial and entrepreneurial roles that pushed the cutting edge, this chapter brings readers unmatched insights: from technology to geopolitics. The two authors have a strong faith in humanity and the power of collective intelligence that pushes technology to new heights. This chapter peeks into the future as air travel begins to turn science fiction into business fact.

1 "Jekyll and Hyde"

Next time you get frustrated by flight delays, it might be good to take a second and reflect on the humbling level of innovation that has enabled you to partake in the miracle of flight.

Aviation has always been defined by pioneering innovation that drove "10×" scale improvements in distance, capacity, speed, and comfort. In 1903, the Wright Brothers, undeterred by many previous rounds of failure, took to the skies for the first powered flight that lasted 59 s and covered a distance of only 260 m. Only 28 years later, Wiley Post successfully circumnavigated the globe in a fixed-wing propeller aircraft, with a total distance of nearly 25,000 km. In the same decade, in 1937, Pan Am commenced regular commercial passenger service across the Pacific Ocean, connecting San Francisco and Hong Kong with a six-stop journey and morphing tiny uninhabited islands in the middle of the Pacific Ocean into refuelling posts for their seaplanes. The Comet—the first commercial jet aircraft—took to the skies in 1949, flying around 50% faster than equivalent piston-engine aircraft of the time. 1969—only two decades later—was quite the year. We "10×'d" the

M. Bastawros
Bristol, UK

O. Haas (✉)
Hong Kong, Hong Kong

T. Carleton et al. (eds.), *Inventing the Almost Impossible*, Future of Business and Finance, https://doi.org/10.1007/978-3-031-36224-8_4

number of people that could cross the Atlantic with the majestic Boeing 747. That same year, we also broke the sound barrier—twice—with supersonic flight on the Concorde; oh, and we also went for a stroll on the moon. It is worth remembering that all of these innovations were paper-driven and developed without the assistance of the fancy computers we rely on today.

By connecting diverse people, cultures, and ideas, commercial aviation has simultaneously made the world smaller and our global economy larger. Planes carried 4.5 billion passengers in 2019 and generated over US$960 billion in GDP—around the size of Indonesia or the Netherlands. The industry is also a major employer, supporting 88 million jobs directly and indirectly, across the globe (ATAG, n.d.). While COVID-19 brought commercial flying to a momentary standstill, air traffic rebounded rapidly once the world's borders started reopening. It is estimated that only 10–20% of the world's population has ever travelled on a plane (IEEP AISBL, 2019), indicating a huge potential for growth ahead.

However, all of this incredible development has come at an environmental cost. 915 million tons of CO_2 was emitted by commercial aircraft in 2019, representing about 2.1% of global human-generated CO_2 emissions and 12% from transportation sources (ATAG, n.d.). If it were a country in 2019, the aviation industry would rank within the top ten emitters globally. Aircraft, of course, also emits a cocktail of additional greenhouse gases (GHGs), including methane (CH_4), water vapour (H_2O) nitrogen oxide and dioxide (NOx), sulphur oxides (SOx), and soot. The fact that these gases are released at high altitudes may increase their global warming effect (IPCC, 1999).

Therefore, the conversation about inventing the future of aviation is inherently a conversation about inventing a more sustainable future of aviation. As we move on from the golden "jet age", the industry's next epoch will be defined by sustainability.

2 Making Flight Great Again

So how do we tackle aviation's environmental issues?

As in other industries, there is ample opportunity to improve the efficiency of existing commercial aviation. Studies have explored how investments in improved air traffic control management, ground operations, and aircraft retrofits can significantly reduce operational emissions (Schäfer et al., 2015). However, in the highly competitive airline industry, these efficiency advancements will likely be passed down as lower ticket prices for consumers, which could trigger a "rebound effect" that leads to even more people flying (Dray & Cebon, 2012).

Given this potential concern, tackling the aviation emissions problem will require addressing the elephant in the room: fuel. Typically, 30–40% of an airline's operational costs are attributed to jet fuel. As a result, the industry has had a strong track record of improving fuel efficiency; aircraft emissions have declined by over 80% since the dawn of the jet-age 1950s (ATAG, 2020). Most of these reductions have been driven by the development of newer, fuel-efficient aircraft. In today's generation, you may have had the chance to travel on a comfortable Airbus A350 or Boeing 787, both of which feature advanced fuel-efficient engines and lightweight carbon fibre bodies. However, these aircraft still fundamentally rely on combusting the same jet fuel that was guzzled in the 1950s—there has not been a significant change in the propulsion technologies of aircraft. The situation is similar to the electronics industry, where improvements in the battery life of your laptop and mobile phone come primarily from higher-efficiency components like microprocessors and screens, rather than significantly better batteries.

The main challenge in finding a zero-emission alternative to jet fuel is weight and specific energy—that is, the amount of energy you can pack within 1 kg of fuel. Planes need to fight gravity to take to the skies, and as such, every gram of weight they carry matters. There simply are not many alternatives for safe, commercialisable fuels that match the specific energy of jet fuel. Though there is no silver bullet, there have been exciting advancements in three key technologies—sustainable aviation fuels, battery-electric, and hydrogen—that, together, can form a solution.

- Sustainable aviation fuels (SAF) are fuels that are made from renewable sources that are processed to have similar characteristics to current jet fuel. This allows them to be distributed through current infrastructure and to be used in existing aircraft. SAF can be made from agricultural biomass, including waste products that do not compete with food crops. Another promising option is Power to Liquid (PtL) which involves synthetically producing SAF from renewable energy and captured carbon molecules. Depending on the feedstock and conversion technology used, SAF may cut lifecycle CO_2 emissions by 30–60% (World Economic Forum, 2020a). Unfortunately, however, SAF still produces contrails (the condensation lines you see streaking across the sky when an aircraft flies overhead), which can contribute significantly to global warming (Pearce, 2019). The fuels also face significant scalability issues—despite growing demand, the planned production capacity for SAF is expected to cover only 1% of global jet fuel demand for 2030 (World Economic Forum, 2020b). Prices are also expected to remain 2–4x higher than traditional jet fuel in the long term (Bain, 2023). Significant investment into new SAF plants, renewable energy, and scaled production will be required to bring these volumes up and prices down.
- Battery-electric—Driven (pun intended) by the growth of the electric vehicle market over the past decade, there has been a significant recent development in battery-powered aircraft. Unfortunately, the Achilles' heel of electric flight is battery weight. Current batteries yield over 40x less energy than the equivalent mass of jet fuel, thereby limiting the speed, capacity, and range of electric aircraft. If charged with renewable electricity, battery-electric planes can achieve the nirvana of zero-emission flight. Also, notable is that electric aircraft are significantly more efficient than conventional or hydrogen-powered aircraft, at a systems level. This has launched a flurry of development in new electric air taxis (often referred to as "eVTOLs") and small aircraft that can enable smaller regional airports to be served more efficiently. If the range issue could be

addressed beyond these applications, electric aircraft could play a key role in decarbonising aviation (Bain, 2023). Beyond improved battery chemistry, there are now exciting developments tackling this very opportunity.

- Hydrogen—There are two primary ways to power an aircraft with hydrogen. For shorter flights, hydrogen fuel cells can deliver power to electric motors with a 75–90% reduction in overall emissions compared to jet fuel. For longer-haul aircraft, hydrogen can also be combusted directly in a turbine, which could lower overall emissions by 50–75% (World Economic Forum, 2020a). A number of prototype aircraft powered by hydrogen have recently taken to the skies, helping to build increased trust around the new technology and quelling previous concerns of exploding Hindenburg airships that may have existed colloquially. The main challenge with hydrogen is the low overall systems efficiency, measured from the energy required to produce the fuel to the energy output by the aircraft's engines. Hydrogen and SAF have a systems efficiency of 20–25%, compared to battery-electric at over 75% (Bain, 2023). This efficiency is important when considering the total energy required to power the bulk of future commercial flights. There will also be high demand for renewable energy from other sectors, so the efficiency of its use matters. There are also challenges in scaling production and developing infrastructure to make clean hydrogen available at a significant number of the world's airports.

Fundamentally, the aviation industry is still viewed as difficult to innovate in—stringent regulation, high research, development and capital costs, long procurement cycles, and highly concentrated competition pose daunting barriers to new innovators. However, as has recently been proven in other similarly challenging industries such as electric vehicles, space travel, and even the meat industry, the right mix of mindset, leadership, and incentives can bring down these barriers and enable a high potential for innovation and meaningful disruption.

There is increased regulatory pressure on the industry, from UN policies to curb aviation emissions to the EU Emissions Trading Scheme (EU ETS) and local regulations in regions like in the Nordics, which are mandating zero-emission domestic flights within the next one and a half decades. This, together with growing demands from corporates and consumers for more sustainable travel, will provide some of the incentives to accelerate and inspire further developments in aviation sustainability. However, despite the growing interest, most of the industry currently agrees that step-change sustainable technologies may only be available in a decade or two. Therefore, it is our conviction that policy is not enough and that more fundamental mindset changes are needed to fuel the invention of the future in aviation.

How can we accelerate developments in sustainable aviation? And how might we innovate to discover other "left field" solutions?

We explore how this acceleration could originate from within existing incumbents or from an entirely new breed of entrepreneurs and innovators that bring a fresh perspective on how we take to the skies.

3 A Deep Look Within

For the last 100 years, aircraft has pretty much looked the same. Yes, aircraft like the 787 has taken carbon fibre to new heights, and the 500-ton A380 has made significant advancements in engineering, but have we really seen seismic change like Tesla and SpaceX?

Considering the key technologies discussed earlier, we would traditionally consider that the fate of the industry today primarily lies with the original equipment manufacturer (OEM). Manufacturing giants Airbus and Boeing currently dominate commercial aviation production, and when they set out to develop new aircraft, they drive an entire ecosystem of suppliers and partners to innovate according to their new specifications. However, are the OEMs currently primed to sufficiently drive the much-needed disruption to shake the whole system up? Maybe not.

To envisage the future of the whole industry, there needs to be also a substantial change within the major OEMs, or at least the ones that want to be part of its future. It will require going back to the drawing board and thinking beyond conventions. Business as usual has no place in the future of aviation. This movement will undoubtedly revolve around the experts, but also untapped talent: the intrapreneurs. Intrapreneurship is a movement that has been gathering significant momentum in many industries, including aerospace. It relies on employees to act like entrepreneurs working on new ideas, but within the organisations they work for. This is vital for the future of the industry because although some intrapreneurial ventures will succeed and others will fail, they will trigger a wave of creativity that could fuel something else. Depending on century-old expertise, it is no longer sufficient if we are to imagine an emission-neutral industry: we need absolutely every possible idea and means out there.

Accepting that intrapreneurs can be also the drivers of a different future requires much more than the talent to get things moving. There must be an ecosystem around them free of politics and agendas, but most importantly an environment where failure is acceptable. It somehow requires a mandate from C-suite to give such disrupters the freedom to constantly challenge the status quo. What this will inevitably trigger is empowerment to try something different without the fear of career suicide. Individuals will no longer try to pick the low-hanging fruit, but will be thirsty to make juice from the whole tree. Once again, is that sufficient? Probably not.

Then comes the innovator's dilemma. Why would Boeing and Airbus invest in new technologies when they are selling the current offering like hotcakes? Well, it is actually a matter of survival. With the world turning against fossil fuels and some countries completely banning their use in the next decade, they actually have no choice.

Airbus' ZeroE initiative has been truly inspiring with ambitious goals of making the first commercial hydrogen-powered aircraft a reality by 2035. However, can Airbus make this happen on its own? Not a chance. They have been very vocal on this actually; they are pushing for as much cross-collaboration as possible. But would Boeing be willing to work with its arch-rival? Debatable. To truly take the industry to new heights, the intrapreneurial model needs to be replicated within every single organisation, but also for such companies to open their doors for cross-intrapreneurial pollination between them. Only then we can expect the cross-industrial team of experts, out-of-the-box thinkers and change-makers to deliver the next breakthrough. Testament to this is the Concorde: a perfect example of multi-state and company collaboration between the UK and France. Undoubtedly, they hit hurdles along the way, and cultural differences probably did not make the programme smooth sailing at the beginning—but in the end, pints and wine glasses were clinked, and the result was a masterpiece of engineering.

The aerospace industry has a history of coming up with new technologies that end up being adopted within other industries (autopilot and anti-collision software to name a few). This is because the amount of money spent in R&D is eye-watering. Airbus and Boeing collectively spend over US$5 billion per year (that is nearly US$100 million per week!) in exploring new technologies that may never see the light of day. We are not even counting the money spent by the engine manufacturers, suppliers, airlines, and governments. But how much of that is duplicated and spent on the same technologies without even knowing? Would not it be better if whatever percentage of duplication is spent together towards a common goal of making the industry emissions-free? Is it not better if the next breakthrough was open source, agnostic to corporate b***s***, and made by organisations, experts, intrapreneurs, and scholars across the world? With the future of the whole planet hanging in the balance, we do not believe that this is the time for protectionism. Purpose-driven and ego-free aerospace companies are the future of a sustainable aviation industry. The capitalist mindset of hunting down every single dollar of profit at the expense of shutting out the competition is not going to work. It may lead to the extinction of the industry as we currently know it.

4 Tony Stark with a Crypto Fortune

What if the next breakthrough did not come from the big players in aerospace? What if the next breakthrough came from an Elon Musk or one of his successors? Elon has been very vocal about tapping into aviation: he already presented his vision of Starship rockets travelling around the globe in the outer space to link cities as far as New York and Shanghai in a matter of minutes. This could bring the industry as we know it to its knees. However, he has often mentioned that he lacks time. Being the CEO of Tesla, CEO/chief engineer of SpaceX, founder of Neuralink, OpenAI, and the Boring company, must keep him a tad busy; so, his attention is elsewhere, *for now*. With the rise of cryptocurrency, there are new members of the billionaire crew who are looking for ways to spend their fortunes. All it might take is a real-world Tony Stark with a passion for flying even half as large as that of the authors of this piece to attract the necessary investment, time, and effort to shape the future of aviation. Will the opportunity to redefine their reputation on the world stage be how these individuals invest their fortunes? They love flying on private jets, so they should want that luxury to exist in two decades' time.

The reality is you do not have to be a billionaire to become the next Elon Musk of aviation. You need vision, the aura to inspire clever people to jump ship from the aerospace giants and tons of perseverance. Take Lilium, for example, the four German chaps who came together at the Technical University of Munich to build a future for urban air mobility or the founders of Zipline who have leapfrogged logistics technology on the African continent into the next decade with a fleet of automated drones. This new generation of entrepreneurs were not born rich, but they were born with passion.

When Facebook announced that it was going to become Meta and redefine itself by investing in the metaverse, it cleverly made a ton of noise. What Meta intended all along is to trigger a wave of new start-ups who will help it accelerate the development of the metaverse (Petaverse is now a thing). Not only this is going to enable Meta to explore new ideas it had not thought of, but also completely buy out the more agile, nimble, and cost-effective startups out there (i.e., the next WhatsApp and Instagram). This is exactly what Airbus' ZeroE programme has done. It has given the world a clear signal that hydrogen was the way forward. As a result, new startups have emerged like Universal Hydrogen and ZeroAvia, which have attracted a lot of attention and cash. Governments across the world from the UK to Egypt are now rushing to commit significant financial resources to enable the industrialisation of hydrogen. The race is on.

In reality, it is unlikely that only one of these aerospace giants, startups, and governments will succeed. Our bets are on mergers and acquisitions, and cross-governmental collaboration within the next decade, enabling a complete shift in the duopoly we are used to today. We expect the Davids ganging up to take on the Goliaths in the short term. The Davids will become so refreshingly nimble that many of the talents within the aerospace giants will want to be part of their mission and innovation culture. However, will the Davids win? Not alone. In the long term, we believe the future of aviation will require the biblical stories to be rewritten: David and Goliath will need to become friends and work as part of the same team.

Take Boeing as an example. The American aerospace giant has been eyeing a way to reinvent itself, and it has been comparatively quiet on its grand vision for sustainable aviation. However, by acquiring eVTOL company Wisk, Boeing has made its short-term strategy clear: de-risk its main business while preparing for future developments like in urban air mobility. As the company works hard to bring back its former glory days, Boeing is buying itself time to figure out which David(s) they want to place their bets on. Some would say that is an astute game to play, we wonder if William Edward Boeing would feel the same. Has the company lost its pioneering spark?

Has the aviation industry lost its appetite to take risks? Is the aviation industry missing a risk-taking Elon Musk to try something radical where failure is deemed acceptable? Absolutely. Aviation requires safety to be put at the fore since billions of human lives are involved, which is not even a question. However, we now live in a world where the evening news shows millions of dollars worth of (uncrewed) prototype rockets blowing up in the name of iterative technology development. Then, a few years later, we see jaw-dropping images of Space X's Falcon 9 booster rockets landing with the perfection of a Swiss watch. Do not get us wrong; we are not advocating for putting passengers on board risky technology and for new developments to bypass rigorous testing and certification. We are advocating to bring back the pioneering spirit of aviation, to try new ideas that do not necessarily fit the mould, and to laugh off mistakes before dusting ourselves off to try again. Exactly like the Wright Brothers did.

5 New Kids on the Block

Wright Brother-level inventors and Elon Musk-flavour entrepreneurship will play a massive role in the future of aviation. It is the visionaries that are going to shake up the conservative industry, push forward the regulators, and progress us to our sustainable future. Encouragingly, there have been hundreds of ambitious new startups and sustainable aviation initiatives that have been founded in the last half-decade. There is fundamental research into improving SAF production and battery chemistry. There are nimble startups developing electric and hydrogen aircraft of various shapes and sizes. Many of these companies have already started flying next-generation aircraft, while traditionalists are still stuck in meeting rooms in endless rounds of discussion. The groundswell is happening and we cannot wait to see how these initiatives shape our future.

But, as innovators, we also not forget to ask ourselves—*what if*?

The future of aviation is not only going to be shaped by the established industry in the West. China is working hard to establish itself as a new aerospace player to be reckoned with—from developing a home-grown space station and new commercial aircraft. China's new C919 jet is technically inferior to its Western competitors, but it has already racked up over 500 local orders. It was developed through a concerted effort coordinated by the government, of course, with some borrowed technology. While this jet currently does not embody any revolutionary sustainability technology, this will quickly change when the government decides that it should. With a few more years of experience, it is inevitable that China is going to shake up the market.

When Russia focuses on the sky again instead of prodding its neighbours, the country also may have a role to play in the future of sustainable aviation. Russian aircraft may not be known for their quality; however, their aerospace industry has historically taken risks that others did not dare to entertain. Russia already flew a hydrogen-powered aircraft in the 1980s—the Tupolev Tu-155. It performed its first flight on the 14th of April 1988, but with the demise of the Soviet Union, all plans were abandoned. Russia's geography, climate, well-established fuel distribution infrastructures, and pocket-filled tycoons would also position it well as a producer of renewable energy and hydrogen.

Russia and China—or any other country that is trying to (re)position itself on the world stage—need to come up with something dramatically *different*. With a more peaceful world order, these countries could position themselves as key enablers of a sustainable future—developing end-to-end clean energy production and supply chains, including designing new generation aircraft and an entire supporting ecosystem. This is a recipe for disruptive change that China is already demonstrating with electric cars and high-speed rail.

6 The Next Messiahs

So, the future of aviation is about sustainability. We have not touched much on other visionary aspects like how air taxis could change the way we travel in cities, how artificial intelligence could transform the way we manage our cluttered airspace, or how the resurgence of supersonic flight could make our world even smaller. That is because, in our view, any one of these innovations would need to be predicated on a foundation of sustainability.

The most exciting prospect for the sustainability challenge is that we have fundamental solutions to the problem on our radar (pun intended). Yes, we still need a ton of hard work to engineer, regulate, and commercialise sustainable aviation fuels, along with battery-electric and hydrogen-powered aircraft; from our experience in the industry, we will not for a second underestimate these efforts. However, from a 35,000-foot view (pun intended, again), we recognise that the challenges at hand are all technically solvable and, with the help of some creative economics, very likely feasible.

However, if big-ticket technical innovations are a decade away from commercialisation, do we keep doing what we are doing and living in denial? Absolutely not. We need to move faster and more creatively, without ever compromising on the fundamentals of safety. Moving faster is not a commercial ambition; in this case, it is about something far more fundamental. Hans Christian Andersen famously said that "To travel is to live". Flying is the most incredible connector of people, culture, and ideas. It is part of the fabric that binds our world together and enables the connections fundamental to human progress. We should continue flying and finding ways for more of us to fly. However, speaking at a tipping point for climate change, we must do so without compromising our planet and the future of our species.

However, we should not *just* be focusing on tomorrow and pursuing just the next big thing. Innovation and change must start today by exploring the smaller, more immediate opportunities. We need to collect and explore every single low-hanging fruit. What is the point in having a zero-emission aircraft by 2035 when every other aspect surrounding that flight is based on technology and operations from the 1990s? From cumbersome transit to and from airports to the overuse of plastics, from renewable energy to sustainable food options onboard. Smaller and more meaningful changes in the 41,000 airports littered across the globe will shift mindsets and prepare the foundations for the major technology breakthroughs to be adopted when ready. As much as the future of the industry is going to be driven by the aerospace giants, the responsibility for change also lies beyond Airbus and Boeing.

Mastering human flight is perhaps one of the oldest and most fundamental of an innovator's dreams. Leonardo Da Vinci's sketches from the 1400s remind us that we have been pursuing this dream for a very long time. Sitting in our $100 low-cost airline seat today, we can only imagine how impossible the prospect of flight must have seemed only a century ago. However, the overwhelming challenge never stopped us—we kept dreaming big and eventually took off for a few hundred metres, then crossed vast oceans, broke the sound barrier, and eventually mastered the art into an efficient commodity. We truly take that for granted sometimes and have managed to come a long way since riding horses and burning blubber to heat our homes.

It took the undying dreams, optimism, persistence, leadership—and sure, a good dose of luck—to achieve every one of aviation's major milestones. And at the heart of these efforts stood remarkable people. These people did not have to be billionaires, they were not necessarily single inventors and they did not even have to be "outside entrepreneurs". Instead, they were fuelled by passion and an out-of-the-box innovator's mindset. While we have put a lot of emphasis on technology in our writing, we should consider that the future of aviation is also fundamentally about people.

When we dare to dream big, we achieve things deemed impossible. We break boundaries, write history, and set new challenges for the future. This chapter was designed to inspire and trigger ideas. Most importantly, it is a call to action: for optimistically visioning the future, for creatively collaborating, and for bringing forward the Messiahs of Aviation who will lead the next movement. We need gravitas and leadership to evangelise and galvanise people, organisations, and governments to work together and achieve the seemingly impossible.

We are secretly hoping that one of those next Messiahs is going to be you.

References

ATAG. (2020). *Benefits beyond borders 2020*. Retrieved from aviationbenefits.org/downloads/aviation-benefits-beyond-borders-2020/

ATAG. (n.d.). *Facts & figures*. Retrieved from www.atag.org/facts-figures

Bain. (2023, May 25). *Will plans to decarbonize the aviation industry fly*? Retrieved from www.bain.com/insights/will-plans-to-decarbonize-the-aviation-industry-fly/

Dray, L. M., & Cebon, D. (2012, December 21). The rebound effect in the aviation sector. *Energy Economics*. Retrieved from www.sciencedirect.com/science/article/abs/pii/S0140988312003428

IEEP AISBL. (2019, October 9). *Linking aviation emissions to climate justice*. Retrieved from ieep.eu/news/linking-aviation-emissions-to-climate-justice/

IPCC. (1999). *Aviation and the global atmosphere*. Retrieved from www.ipcc.ch/report/aviation-and-the-global-atmosphere-2/

Pearce, F. (2019, July 18). How airplane contrails are helping make the planet warmer. *Yale E360*. Retrieved from e360.yale.edu/features/how-airplane-contrails-are-helping-make-the-planet-warmer

Schäfer, A. W., Pilling, C., Wadud, Z., Pangbourne, K., & Cebon, D. (2015, November 23). Costs of mitigating CO_2 emissions from passenger aircraft. *Nature News*. Retrieved from www.nature.com/articles/nclimate2865

World Economic Forum. (2020a, November 11). *Clean skies for tomorrow: Sustainable aviation fuels as a pathway to net-zero aviation*. Retrieved from www.weforum.org/reports/clean-skies-for-tomorrow-sustainable-aviation-fuels-as-a-pathway-to-net-zero-aviation/

World Economic Forum. (2020b, November). *All aircraft could fly on sustainable fuel by 2030, says World Economic Forum report*. Retrieved from www.weforum.org/press/2020/11/all-aircraft-could-fly-on-sustainable-fuel-by-2030-says-world-economic-forum-report/

Mina Bastawros is the Vice President of Creative and Digital Marketing at Airbus, where he has held multiple roles across the last decade from engineering to procurement. He was hand-picked by Airbus' CEO to lead a moonshot team that would build a path for the company to win the future. During that journey, he discovered his skills as a storyteller, disrupter, and intrapreneur. He holds multiple degrees in aerospace and aeronautics. When not surrounded by aircraft, he is an advisory board member of startups, institutions, and also the Head of Marketing & Communications for the Deep Space Initiative. Mina often gives talks and brainstorms with other Fortune 100 companies to inspire future disrupters and spread the mindset of being a pirate in the navy. In his free time, Mina unwinds through street photography with the aim of giving a voice to the voiceless and those forgotten by society.

Oliver 'Ollie' Haas is the Head of Partnerships and Strategy at Magpie Aviation, a San Francisco-based venture enabling long-range, zero-emission flight. Previously, he was a founding team member of Climate Impact X, a global marketplace for high-quality carbon credits based in Singapore. At Cathay Pacific, one of the world's leading airlines based in Hong Kong, Ollie led innovation teams developing everything from the passenger mobile app to customer service AI assistants and a major customer experience program for the future flagship 777X aircraft. He helped to transform the thriving innovation consultancy PALO IT into Asia's first technology B-Corp. While at LittleBonsai, the sustainable design studio he co-founded, Ollie developed the world's most talented keychain and tried to make a better toothbrush. Ollie studied engineering and sustainable design at Olin College and completed a professional graduate program in innovation and environmental studies at Stanford University.

Creating a Work Culture for Team Innovation and Imagination

Louise Kyhl Triolo and Tamara Carleton

Abstract How do you create the conditions for innovation within an organization to truly unleash the imagination of teams? This chapter considers the real-world human factors and conditions that our experience has shown as critical for lasting positive shifts to happen in organizational culture. We present an organizational model for shifting innovation culture, highlighting several best practices that we have learned—sometimes the hard way—as well as practices underway at other notable companies like global furniture company IKEA.

How do you create the conditions for innovation within an organization to truly unleash the imagination of teams? Much has been written about the topic of culture change (and corporate change in general), so our thoughts are not intended as a meta-analysis of existing models and methods from management and organizational behavior studies. Instead, this chapter considers the real-world human factors and conditions that our experience has shown as critical for lasting positive shifts to happen. We present an organizational model for shifting innovation culture, highlighting several best practices that we have learned—sometimes the hard way—as well as practices underway at other notable companies like global furniture company IKEA.

Organizational change is directly related to inventing the future, even if not necessarily recognized as such by senior management and human resources (HR) departments. If you are not creating a new future state, then what is organizational change for? Leaders and teams tasked with inventing the future face a parallel challenge of changing the culture and mindsets of people around them to accept what is new and different to the current business. A new business model or game-changing solution rarely succeeds under the former organizational culture and talent model.

L. Kyhl Triolo
Heidrick & Struggles, Paris, France
e-mail: lkyhltriolo@heidrick.com

T. Carleton (✉)
Department of Mechanical Engineering, Blekinge Institute of Technology, Karlskrona, Sweden
e-mail: tamara.carleton@bth.se

T. Carleton et al. (eds.), *Inventing the Almost Impossible*, Future of Business and Finance, https://doi.org/10.1007/978-3-031-36224-8_5

We believe that when an organization is serious about its business transformation, which typically entails a massive changeover in ways of working together, then the entire change is—or should be—geared toward inventing its own future state.

1 Defining an Innovative Work Culture

What does "work culture" really mean? Simply put, culture is the sum of how things get done in a social environment. In a company, this culture is reflected in how employees and customers behave and are treated. Culture is further embodied and perpetuated by senior leadership, influencing who gets hired, who is kept, and what motivates everyone to do their best (or worst) performance at work. Work culture often gets discussed as a competitive lever in the business press, and the benefits are numerous. Research studies show that work culture directly affects employees from internal measures like job performance and motivation to external factors like customer satisfaction and service quality.

Beyond these formal measures, culture reinforces the informal expectations for how people should behave and interact together. Think of culture as your organization's climate. Like patterns of weather over a long period, conditions at work build the enduring shared environment for your teams. Culture becomes the collective attitudes, beliefs, values, assumptions, and behaviors across the workforce. And while humans may have limited success controlling the weather in the sky, company leaders and team managers can directly change the atmospheric conditions for their people.

A healthy work culture leads to employees who are more creative, contribute more, stay longer, and build better relationships with coworkers and the broader ecosystem. A healthy work culture also breaks down boundaries between siloed groups, reduces office politics, and improves workflow overall, while an unhealthy work culture has a detrimental effect on team performance and the human experience. As importantly, a healthy work culture generates a work environment that values and supports innovation, so that teams can actually make innovation happen. Imagination then becomes the innovation power to continually see new opportunities and introduce different ways of thinking and doing.

2 A Model for Shifting Organizational Culture

Why is work culture so important to innovation? Culture influences how things get done—for example, how teams introduce new ideas or take an entrepreneurial hustle to a new program. Culture can also work against new business strategy; when not aligned, people will rely on the status quo and mismatched metrics that delay or obstruct genuine change forward. As such, any organization faces difficulty when changing its work culture. Generally, large organizations must overcome decades of legacy with firmly entrenched ways of working, and many startups are led by at

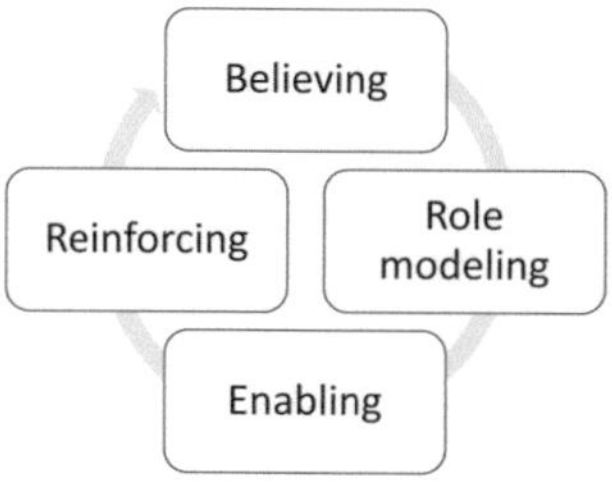

Fig. 1 The Kyhl Triolo model for organizational change. Drawing by author

least one strong-willed or charismatic founder who sets a dominant organizational tone. Changing the culture within any organization, regardless of its size, can become overly complex and even onerous to do because it involves the simultaneous tasks of shifting mindsets, behaviors, practices, processes, and platforms—all of which can be a daunting task to take on.

At the core, we believe the change becomes simpler when you focus on humanizing the change. By humanizing, we mean putting the person at the center of every process or conversation. In practice, this becomes a set of small actions that add up, such as directly acknowledging the emotions in any new process request, creating immersive experiences, talking face-to-face whenever possible, and using understandable terms that any layperson can grasp. In doing these types of actions, both teams and individuals go from feeling that change is happening to them to being a part of the change. Culture change is then a movement, not a mandate.

Another basic truth is that work changes when you change work. In other words, leaders must show courage in going through with any change they promise. American philosopher-architect Buckminster Fuller once wrote, "You never change something by fighting the existing reality. To change something, build a new model that makes the existing model obsolete." His quote is a recognition that dabbling around the edges of change does not lead to lasting change; instead, leaders must take a systemic view that shows the old ways of working are no longer as relevant or purposeful as they visibly embrace new ways of working.

Based on this systemic approach, this chapter presents a model of organizational change with four related dimensions of believing, role modeling, enabling, and reinforcing innovation culture change. Louise Kyhl Triolo has used and refined this model over the last decade to guide inventing a future innovative work culture. See Fig. 1 for the Kyhl Triolo model for shifting organizational change.

3 Dimension of Believing

The first dimension in the model is believing. Change ideally begins with the belief that a group must change and that the change is possible. Believing links the desired mindset and values of innovation with the target behaviors, and these behaviors are then often encapsulated in a change story or change narrative. What leadership believes determines what they think is doable. As a sadly common example, if top management believes they lack the right innovative talent and that no internal team is good enough, then they will create the systems, processes, and programs to correct this gap and then focus on evaluating, often through performance management, and replacing talent rather than developing existing talent. This response does not deeply address the core mindset and underlying beliefs needed for thorough cultural change.

Bob Sternfels, who took over strategy consultancy McKinsey in 2021, saw this issue firsthand. He said, "One of the biggest pivots we can make in our own talent-development processes is pivoting from a culture that's obsessed with evaluating people to one that's developing people" (Cutter, 2022, para. 20). In his first months, Sternfels embarked on a listening tour, meeting with over 300 chief executives and talking with McKinsey employees worldwide. Part of the change he initiated was to support employee growth. He added, "And we're wiring in a whole bunch of changes around that to try and get people back to [doing] what you want to do as a professional, not what people are telling you in terms of the evaluation process" (ibid.).

To truly change, you need the hearts and minds of employees to match what leadership believes. For many people, believing is both seeing and feeling genuine change is happening. The messages of improvement cannot only be communicated top down from senior management; instead, the desire to transform must be experienced from the inside out. Consider Airbus Helicopters (AH), a division of Airbus Group and also the world's largest helicopter manufacturer. In the mid-2000s, AH started a corporate transformation journey by inviting staff to see how the existing work culture was blocking innovation. An initial diagnostic assessed internal culture, showing multiple obstacles based on industry data and employee input. For example, data showed that in recent years AH had dropped from being #2 to #5 in customer satisfaction compared to competitors. In response, an AH task force created several "transformation rooms" onsite at their European offices as physical places and virtually shared with other AH offices where this data with examples were shown to staff in full transparency. All AH managers and employees could symbolically look in the mirror, which made the need for change clear. As a result, more employees understood the current situation based on real evidence, saw management was serious and open about changing from the past, and felt they could be co-creators in the change ahead.

This is where a change story becomes important for everyone to hear and tell within their teams—not just at the senior level. A change story is a short explanation, verbal and/or written and sometimes visually shown, that summarizes what the change is, why the change is needed, and how employees can start the change. A change story helps a team or employee see how the new way(s) of working relate to the organization's overall vision. The core element of believing is creating a story to inspire or aspire to, what you might call a "burning imperative." What is the change we believe will make the difference? In this sense, a story is meant as a simple narrative of the change a group wants to see, ideally longer than a tagline yet less than a multi-page scenario. These stories should also capture the language and tone that a group wishes to emulate. At AH, the process entailed creating a series of mini-stories that described various change moments 3 years in the future. One mini-story was that a newcomer would know an AH customer immediately. The team also worked to capture a story of future change in 5 years that described how a specific cultural shift looked and felt like, with an emphasis on desired behaviors, such as service mindedness and problem solving. These stories were then posted back in the transformation rooms for all employees to see and react to.

4 Dimension of Role Modeling

Senior leaders then act on these new beliefs through the second dimension of role modeling, which requires them to step forward and make the target behaviors visible and embedded into the desired ways of working. In doing so, others around them become empowered to act the same. This creates a constructive chain reaction of behavioral change from senior leadership to middle management down to junior staff. From experience and research, leadership behaviors are the primary mechanism that reinforce cultural habits because teams and employees generally follow management's example.

This step of role modeling is especially hard for top management to embrace. To transform means to develop into something entirely new, to shed what is unnecessary and no longer useful. More than just to speak to the desired change, each leader must develop his or her own sense of personal change needed to embrace the wider organizational change. Sometimes, it requires a complete change in management to shift an organization's culture, as the leaders who have been in established roles or legacy industries for a long time have created that same culture they are now responsible for changing. So naturally, it will be hard for them to imagine new ways—or consider what they should change next for themselves personally—to set a different tone for those around them.

In most situations, we have found that change is best role-modeled from the top down first and then from the middle out and bottom up in parallel. This role modeling takes time to do. In one manufacturing company we know, a task force created a 9-month transformation program, starting with the senior leaders. This new leadership program focused on building experiential self-awareness alongside new practices and advanced leadership skills, so that the leaders could form new habits related to innovation. Working in small cross-functional groups of 15 participants or less, leaders could feel comfortable yet have enough peers to practice the new behaviors and practices together. The cross-functional mixing was deliberate to help break down internal silos and spark more relationship building across divisions. As the program progressed, each cohort had senior mentors assigned from the previous cohort to ensure continuous leadership involvement and consistency of messaging. Over 3 years, over 2500 managers were trained via these small cohorts, and the alums helped to launch the next sessions, further reinforcing the importance of themselves as role models.

A useful tool at this step is finding the "critical few," adapted from Florida-based consultancy Katzenbach Institute. The tool is based on the belief that you have all you need in your work culture; it is just not evenly distributed. Which critical few behaviors and practices would be most impactful in shifting the culture if more people adopted them? Some people are already modeling the innovative change a company seeks through their behaviors and actions. Recognizing and highlighting these individuals will help others see that the desired change is already happening around them. This concept is similar to finding the "bright spots," described by the Heath brothers in their book *Switch* (Heath & Heath, 2010).

5 Dimension of Enabling

However, a singular focus on developing leadership is not enough to accelerate change and make it stick within an organization. The third dimension expands role modeling from leadership to positively enabling broader change adoption across all teams and employees. In any shift towards transforming into something new, learning is required, which is supported through ongoing experimentation and practice.

A spectrum of enabling exists from simple to complex. As a simple example, in organizations desiring to change from a more bureaucratic formal system to a more intrapreneurial system, a small shift in how the governance systems are named can trigger positive change and allow for freer and more creative thinking to manifest. When creating the governance system for "Dream Big," which was a major culture shift initiative engaging Airbus Group's global workforce to imagine bold new futures, the project task force created a Guardians Committee instead of the usual Steering Committee. The objective was to remind the executives on the committee about the importance of serving as protectors for new ideas and innovative talent, which reinforced new role modeling behaviors for leadership. In addition, this simple naming shift from Steering to Guardian helped create the expectation of a positive, direct, and authentic dialogue between leadership and teams. While a name change may seem silly at first glance, studies show that language creates a powerful mental model that sets the tone for certain behaviors, so by adopting the desired language of innovation intentionally, leaders can often prompt the corresponding interactions.

More broadly as part of enabling innovative change, internal communication and messaging play a crucial role in how people experience an organization's culture. Expectations are set through the information that is shared (or not shared), the symbols that are displayed, and the language and tone used for messaging. All this communication tells employees what is important (or not important) in an organization and how to navigate old or new culture. Acknowledging and being able to shift these elements will influence people's experience and provide new guidelines and pointers for what is accepted or not accepted in a given culture.

Consider this example from a Silicon Valley tech company, whose leadership somewhat addressed the first two dimensions in the change process, yet regularly failed at the third dimension for truly enabling change. When this company had set out to transform in prior years, efforts stalled regularly. The company's CEO upheld a parent-child relationship towards his executive team and more broadly to all employees. This stance was displayed through internal communication mechanisms, such as occasions where he would "teach" staff about his new strategy—symbolically kept secret for months beforehand even to his closest executive team—using highly scripted annual internal keynotes. This approach served to spotlight himself as the primary role model and further perpetuated the idea that perfection equals success. Staff perceived the language and tone deployed as "follow me; I know best," which set an expectation to obey "senior orders." These messages had a strong undertone of avoiding any confrontation. The combination of top-down orders with non-confrontational beliefs created an internal culture that quelled any team imagination because no one dared to speak up or try new ideas, essential to innovation. In this environment, the corporate communications and HR groups struggled to enable any genuine organizational change and shifts in employee behavior, and subsequently, staff turnover was high.

During the enabling phase, a corporate communication team can help accelerate the transformation to change. In our example from Airbus Helicopters, by focusing on when, how, and what information was provided to internal teams, the task force discovered that communication from top management seldom reached front-line employees. Although the C-suite believed all staff levels were aware of the company's new direction and understood the consequences on their jobs, in actuality, over half of the workforce was completely oblivious—rendering any change effort as meaningless. Outdated internal communication flows were preventing the broader effort to create a culture of transparency and trust, which are foundations for an innovative culture to flourish. In response, the task force quickly established new forms and rhythms of internal communication, such as weekly e-newsletters sent directly to individual mailboxes, monthly CEO calls, and new work practices to prompt leaders, including the CEO, to continuously "walk the floor" and interact personally with front-line teams.

Enabling new ways of working does not happen instantly; new group processes and new team habits require ongoing learning and practice through experimentation. Although leadership sets the tone and shows the desired example, teams and employees need support in how to approach their job differently and apply new mindsets, tools, methods, practices, and behaviors. During its company transformation, AH invested in a completely new learning academy to train employees in the new techniques needed to become more innovative. Employees directly facilitated multiple team learning opportunities in order to learn by teaching through a peer-to-peer approach. By listening, talking to, and seeing what fellow colleagues were doing, plus encouraging each other to try experimenting—even failing—in a shared and supportive environment, overall learning became much quicker, and changes were implemented with a higher success rate to operations, manufacturing processes, and customer relationships.

6 Dimension of Reinforcing

The final dimension in the Kyhl Triolo change model is reinforcing, as organizational systems become secondary mechanisms that reinforce culture. These systems span formal platforms, processes, and procedures to informal rites and rituals of teams within an organization. Ultimately, the dimension of reinforcing is the one dimension that will make a transformation endure. Think of reinforcing as the infrastructure needed for an organization to adapt faster and with more resilience to any future disruption or change. If infrastructure is outdated, inflexible, and even self-destructive, no lasting positive change can happen. Efforts in this dimension take time and patience to do!

In the examples mentioned of AH and Airbus' Dream Big program, infrastructure surgery focused on people-related systems, which reinforced overall behaviors and beliefs on what was rewarded or who was promoted. Within AH, a complete remake was undertaken of all essential HR processes—including new hire onboarding, performance management, talent acquisition and management, staff and leadership development, and compensation—to fit with the desired new innovative culture. All AH leaders and managers were trained with real-time simulations and role plays to detect, select, and evaluate current and new employees based on the innovative mindset and more inclusive behaviors desired for the organization. The onboarding process was co-developed with key groups outside HR to develop a more agile team culture, and systemic actions included communicating upfront with new hires, ensuring tech equipment on day one, creating a buddying system for peer-to-peer learning, and storytelling about the company history. One key desired behavior was to treat every new employee as the company would treat a new customer by engaging in proper feedback loops and ongoing care. AH's goal-setting process and performance management system for employees were both revamped, inspired by Toyota's Hoshin Kanri process. A critical change was to shift from purely individually set goals to start with team-defined goals, which included new processes to encourage teams to recheck their goals with internal customers.

Major shifts in the management-employee dialogue included establishing new ways of setting goals as a team rather than only top-down directives and creating two-way developmental conversation around goals and expectations with more emphasis on future possibilities and less emphasis on performance evaluation. Another system change was in how employee performance bonuses were calculated, aiming to ensure a more objective, team-, and company-oriented evaluation of outcomes. Company objectives and evaluation percentages were also changed to favor future longer-term strategic outcomes, and these metrics were displayed, discussed, and shared with all employees through a team-oriented, objective-setting process. Together, these systemic shifts helped dissolve the previous organizational culture of secrecy and inefficiency and reinforced a new, healthier team-based culture for innovation at AH.

7 Innovative Culture Change at IKEA

IKEA offers an excellent case study of innovative organizational change, which we describe through the lens of the Kyhl Triolo model. Started in 1943, IKEA has become the world's largest furniture retailer, selling ready-to-assemble furniture, kitchen appliances, and other home services worldwide. In terms of organizational structure, Ingka Group is a separate entity that operates IKEA Retail as its core business, which include 465 IKEA stores, shops, and planning studios in 32 countries (Ingka Group, 2023). Although Ingka Group is one of 12 franchisees for IKEA, the group generates 89% of all IKEA sales worldwide, so Ingka Group serves an important role in instilling IKEA's core values and culture more broadly across the entire IKEA ecosystem (ibid.).

Emerging from the COVID-19 pandemic, Ingka Group wanted to further strengthen its internal core values and empower IKEA colleagues. In 2021, its global HR team addressed the model's first dimension of believing by introducing a new shift from "leadership by the few" to "leadership by all." Alejandra Piñol, Ingka's deputy people & culture and business partnering manager, and previous group talent manager, explained: "The starting point was our core belief that everyone can lead; no exception. Leadership by all defines the most critical aspects of leadership that we will focus on to truly develop our business and ourselves" (personal communication, August 19, 2022). For Ingka Group, the premise of leadership-by-all was based on the notion that "I make a difference," in which every individual contributes to company growth and innovation. Piñol said, "When we empower each of us to lead, we become faster, and more agile and innovative in responding to customer needs" (ibid.).

As part of role modeling change for leadership, which is the model's second dimension, Ingka Group introduced seven new leadership expectations. Piñol said, "We did not want our leadership idea to be seen as a campaign or a one-off initiative. To really trigger behavioral change and involve all co-workers in making the change happen, we wanted to create a movement that would permeate through the entire organization and be driven by the company's leadership, the 21,000+ managers, and 170,000 co-workers." The team also took a broader view of role modeling. The seven leadership expectations, which replaced six capabilities that had been applied only to IKEA managers, now pertained to all 174,000+ employees. These seven expectations for leadership are as follows:

1. I create results.
2. I collaborate and co-create.
3. I develop myself and others.
4. I communicate with impact.
5. I navigate the unknown.
6. I lead by example.
7. I develop the business.

As part of enabling this cultural change, which is the third dimension in the model, the Ingka team recognized the importance of practice and experimentation. Piñol described her team's approach, saying: "Managers need to adopt a 'continuous growth' mindset: think about what aspects of their leadership style may be obsolete and how they can develop themselves to become better leaders, helping coworkers reach their fullest potential." In the first year, she acknowledged that the company was in the early stages of this change. Their starting focus was to "create awareness and deepen the understanding of why we are doing this, and what is required from our managers," a role that the company will continue to evolve with manager input. In addition, Ingka's team introduced a compass as a simple visual metaphor that symbolizes the clear direction and shared navigation across the organization. This metaphor acts as shared vocabulary that helps everyone develop a new way of seeing the change together. Piñol explained that "the compass is a ubiquitous symbol present in all our communications; it reinforces our key message and represents the seven leadership expectations."

Inkga Group has relied on a mix of outreach, training, and learning tools to reach its IKEA network of local stores, shops, and planning studios. The local teams are then responsible to implement and bring the seven leadership expectations and "I make a difference" message to life. The internal reception has been positive. Ingka has held over 400 sessions globally for a communication workshop, presenting the concept framework and related tools. Over 16,000 IKEA colleagues have taken a virtual learning introduction, making it the second most consumed, non-mandatory training within Ingka Global in FY22. Moreover, 7 months after the initiative started, Ingka found that 24 out of 26 IKEA markets, or 92%, have introduced these concepts within their own organization and in several global functions. Each market has adapted the concept in their own way; for example, IKEA stores in Portugal and Norway held a "leadership week," while Germany organized drop-in sessions to discuss the new leadership behaviors. Piñol noted that countries are currently in the process of translating and adapting according to their needs.

As the last model dimension of reinforcing change, Ingka's People and Culture team have begun to ensure all supporting organizational systems bolster the new values. For example, they launched a related internal initiative in FY2023 for a new performance management concept called Performance Success that is rooted in the new leadership approach. The premise is for IKEA coworkers to take ownership of their own performance and development, while supported and coached by managers along the way. Another effort is to relay how everyone across the IKEA ecosystem builds shared leadership together. Piñol said, "Storytelling is an effective way to explore the concept, which is why we are now seeking real stories from throughout the organization about challenges, learnings, accomplishments, and personal growth, both from an individual and team standpoint."

8 Conclusion

The main challenge in building a corporate culture of innovation is the continuous effort it takes to develop and maintain the organizational readiness to embrace uncertainty and the new ways of working. The Kyhl Triolo model for organizational change addresses four related dimensions towards creating an innovative culture: (1) identifying and communicating the change in beliefs from the prior state to a new desired state; (2) role modeling the new change with leadership visibly and consistently; (3) enabling and investing in the new mindsets, skills, behaviors, and practices needed for an intrapreneurial organization to thrive through ongoing training and learning by doing; and (4) reinforcing and aligning the overall transformation with the original purpose for change across all organizational tools and processes. Change does not stop at the first or second dimension. Altogether, the four dimensions of believing, role modeling, enabling, and reinforcing a new cultural change offer a real-world framework that creates the lasting conditions for teams in unleashing imagination.

References

Cutter, C. (2022, January 8). McKinsey's top executive wants to change how the firm operates. *Wall Street Journal.*

Heath, C., & Heath, D. (2010). Switch: How to change things when change is hard. *Crown Business.*

Ingka Group. (2023). *Ingka Group governance.* https://www.ingka.com/this-is-ingka-group/how-we-are-organised/

Louise Kyhl Triolo is a principal in Heidrick and Struggles' Paris office. Louise's passion lies at the intersection of leadership, culture, innovation, and sustainability. She leverages her skills to support organizations to become more agile and intrapreneurial.

Tamara Carleton is a globally recognized expert in radical innovation. As founder and CEO of Innovation Leadership Group, Tamara works with business and government leadership teams to help them build innovation capability and lay out bold futures. With a penchant for creating new methods and tools, Tamara is also a renowned educator, teaching strategic foresight and innovation at multiple universities worldwide. She holds a doctorate in mechanical engineering from Stanford University, where she explored DARPA's enduring innovation practices. She is the author of *Building Moonshots: 50+ Ways to Turn Radical Ideas into Reality* and the *Playbook for Strategic Foresight and Innovation.*

Dealing with Complexity in Uncertain Environments: Wargaming in Transition

Daniel F. Oriesek and David T. Song

Abstract In times of uncertainty and high complexity, business wargaming provides an opportunity to improve decision-making by focusing on the key drivers of developments and practically exploring the future in an experience-based way. The method is in essence a qualitative approach, ideally augmented with quantitative control variables. In such a combination, business wargaming will not only capture the human dynamics of competition but also describe the resulting effects of decisions quantitatively. The aim of this chapter is to provide the reader with an overview of the current state and recent developments in the field of business wargaming while taking a cautious look at what may lie ahead given the recent advances in technology. Furthermore, this chapter provides insights into practical aspects of game design, the selection, and background of participants, the importance of moderation and coaching, the potential of technology, as well as the need for good control models and an after-action review process.

With about 1.5 billion units sold annually, the Apple iPhone not only has been a real success story for Apple but has also revolutionized the way we interact with our cell phones today. Moving beyond simple voice communication, the "smartphone" has changed the way we interact, watch television, work out, order food, tune our guitars, perform banking transactions, orient ourselves, and date. There are hardly any limits to the types of applications that run on a smartphone today. By 2022, about 6.6 billion smartphone subscriptions existed in the world. Based on this figure, about 80% of the global population would have access to a smartphone, not taking into account differences in penetration rates throughout the world.

Although in the early 2000s, other multifunctional devices existed, i.e., combining a cell phone with a PDA, email, and rudimentary web functionality, it was not

D. F. Oriesek (✉)
Bern, Switzerland
e-mail: daniel@oriesek.com

D. T. Song
Stanford University, Stanford, CA, USA
e-mail: dtsong@alumni.stanford.edu

T. Carleton et al. (eds.), *Inventing the Almost Impossible*, Future of Business and Finance, https://doi.org/10.1007/978-3-031-36224-8_6

until the launch of the original iPhone 1 on June 29, 2007, that we can really begin to speak of what is today commonly referred to as the "smartphone."

The credit goes to Apple for being the first to turn the possibilities of mobile data into a final, integrated product, introducing key innovations and integrating the iPhone into a whole universe of music, while opening the platform for developers to create a wealth of applications that have turned smartphones into the indispensable staple they have become today.

While all of this is common knowledge today, few people know that prior to the introduction of the iPhone 1, a series of wargaming exercises took place with commercial companies, such as telecommunications providers, in order to explore the possibilities of mobile data applications or the future of the music industry. Derived from these client wargames in the industry, a scaled-down, simplified version, without all the confidential elements, was conducted with students across top business schools in the Unites States and in Europe.

1 The Booz Allen CEO Challenge

This event was launched in 2005 on location at Harvard Business School and the Kellogg School of Management. Each session involved around 30 students from 10 to 15 top business schools. The case to be explored focused on the development of the portable audio device industry in the United States, involving questions as to what devices would look like; what functionality they would have; whether portable audio players would eventually merge with cell phones; and what role the possibilities of mobile data (i.e., Internet on the go) would potentially play. The event revolved around business wargaming, designed in such a way that it could be played with business students without lengthy preparation or the need for particular industry expertise. The event lasted for two-and-a-half days and started with an introduction to strategy formulation (something students had previously heard about) and business wargaming as a strategy tool. What followed was a crash course in the industry (i.e., mobile data or portable audio devices), with information on the most important trends and players, followed by detailed instructions on how to play the game and how the teams would be divided up. While the competitor teams had the task of analyzing the information and coming up with specific offerings (product, price, promotion, distribution, and services) and an overall strategy for their company, the market and control teams had the task to develop a market view and keep everyone honest and within the possibilities of real life. Over three so-called moves, the competitor teams fought over market shares and tried to offer innovative products and services in order to outsmart the competition while profitably serving to change customer needs. At the end of each move, the competitor teams received qualitative and quantitative feedback from the market and control teams. They received detailed market share numbers, approximate profit margins, and relative change information. The qualitative feedback consisted of statements from the various customer segments: what they liked or did not like about specific offerings.

At the end of the event, students were asked to reflect once more on what they had experienced and what they had learned. This synthesis of views yielded interesting insights, and the teams were able to predict a number of the developments that would later follow in real life over the next couple of years. Case in point was that at least one team per event (this particular game was played with more than ten sets of different students) always anticipated the introduction of an "iPhone" with some of its key features before it actually happened almost 2 years later. They predicted a fully integrated device (phone, music player, navigation system, videoconference tool, wallet, health device, pocket computer) with an open platform that would drive a multitude of potential applications. Findings that are more general included that the industry would move toward hybrid devices and that the biggest threat for Apple and its iPod would come from the mobile phone manufacturers if only they could get their products together. The rest is history.

In times of uncertainty and situations of high complexity, gaming and particularly wargaming provide an opportunity to reduce complexity by focusing not on everything, but only on key developments and practically exploring the future in an experience-based way.

2 What Is Wargaming?

Wargaming, also called war games, originates in the military. The aim was to prepare commanders and staff officers for unforeseen developments on the battlefield and thus give them an advantage. The actual inventors were supposedly the Chinese, who began to carry out such simulations about 5000 years ago. Wargaming was further developed in the Prussian Army and after the Second World War at the US Naval War College. Today, wargaming is still used by militaries around the world and for a wide variety of issues.

In the business environment, the methodology of wargaming appeared in the 1950s when the American Management Association first used the method for training purposes. From the mid-1980s onwards, actual business wargaming was increasingly used in companies for specific, business-related questions and was developed into an independent methodology.

Business wargaming, which Daniel Oriesek and Jan Oliver Schwarz describe in their book, first published in English in 2008, is not a standard product that can be used across industries and possibly in a slightly adapted form repeatedly. Rather, it is a method for dynamically simulating decisions and illustrating the associated consequences with the aim of drawing the right conclusions and defining specific measures for implementation. Business wargaming is individually composed and structured for each company and each question. Business wargaming is, therefore, a qualitative approach, supplemented by quantitative control variables, which is able to map the human dynamics of competition, but also to systematically capture the resulting effects, and also to describe them quantitatively quite accurately. Besides the fact that human dynamics cannot be represented by sterile test scenarios or computer simulations, the participants of business wargaming go through an intensive, interactive experience. This is very close to reality and allows experimentation with approaches and alternative forms of collaboration. In business wargaming, managers take on the roles of their own company, customers, competitors, and other relevant stakeholders. They analyze and virtually experience how the environment develops and whether scenarios previously assumed to be probable would actually occur or not. By putting themselves in the shoes of their market and competitors, managers can overcome certain operational blindness and understand which assumptions about the environment, competitors, customers, etc. are likely to be correct and which are more likely not. Intensive experience also improves understanding in the overall organizational context, and as a result, decisions can usually be made faster, better, and, above all, more closely coordinated as a team after business wargaming. Thus, in particular, internal blockades due to different perspectives (silo thinking) are specifically reduced, and the performance of teams across the strategic, operational, and tactical levels is improved.

3 Success Factors for Playing (War)games and Turning Results into Action

The most important aspect of the design phase is to determine the objectives of the wargame clearly and, more specifically, distill the key question the team will need to have found answers to by the end of the game. Some of the key questions for game design are:

- What are the key questions the wargame needs to answer?
- What are the most relevant yet unaddressed issues in the organization?
- Who should participate in the process during the design and play stages? (From which business units at what level and what should be their role in the exercise, i.e., experts vs. participants?)
- What is the focus of the game? Rather than trying to model every detail of the business, what products, stakeholders, competitors, customer segments, and markets are most relevant in order to answer the set of questions?
- What should the time horizon of the wargame be?

Answering these questions thoroughly and gaining a consensus view at the outset will help to stay focused and arrive at good results.

When designing games, four general design principles apply, which were described by McGonigal (2011) as follows:

- The **goal** is the specific outcome that players will work to achieve. It focuses their attention and continually orients their participation throughout the game. The goal provides players with a sense of purpose.
- The **rules** place limitations on how players achieve the goal. By removing or limiting the obvious ways of getting to the goal, the rules push players to explore

previously uncharted possibility spaces. They unleash creativity and foster strategic thinking.

- The **feedback system** tells players how close they are to achieving the goal. It can take the form of points, levels, a score, or a progress bar. Alternatively, in its most basic form, the feedback system can be as simple as the players' knowledge of an objective outcome: "The game is over when…" Real-time feedback serves as a promise to the players that the goal is definitely achievable; it provides motivation to keep playing.
- **Voluntary participation** requires that everyone who is playing the game knowingly and willingly accept the goal, the rules, and the feedback. Knowingness establishes common ground for multiple people to play together. Moreover, the freedom to enter or leave a game will ensure that intentionally stressful and challenging work is experienced as a safe and pleasurable activity.

In addition, the practice has shown that two additional factors play a role in maximizing the success of wargames:

- **Introduce participants** to their roles. Ideally, this should happen prior to a wargame, e.g., the evening before, to enable participants to immerse themselves in their role. This can be supported, for instance, by exercises that require the groups to think themselves into their role.
- **Surround participants** during the wargame with as many artifacts of the assigned players as possible. This can range from logos, marketing material, videos, and products to hats or t-shirts participants are wearing.

Once the design parameters have been set, the requisite information about the main competitors, the market(s), customers, current and pending regulation, technology trends, possible unexpected events, and so on need to be collected, which are then distilled into the gamebooks (basically the screenplays describing how to play the game) and into the market and control models, which are designed to provide some objective measures to key parameters for the game. The game books contain information about the client company and the competitors in the game, a market overview, as well as the most important trends, regulations, and technological developments in the industry.

In order to get the most out of the game, the participants, regardless of their team, need to receive a proper briefing. This briefing, which can make up the last step of the preparation or the first step during the game's execution, will set the common ground for all parties involved. During the briefing, the teams receive a crash course in strategy and wargaming, typically including a high-level summary of industry trends and facts, which may include a detailed competitor briefing, as well as a detailed explanation of how to play the game. During the "how to play the game" part of the session, participants are divided into various teams, assigned coaches, and familiarized with the expected deliverables and how to communicate with other teams and the market, and given a very detailed schedule.

Typically, three moves are played out and the teams are busy trying to meet their schedule. First, although it is called a game, this is serious business, and it is of utmost importance to the success of the exercise that every individual involved gives the game their best shot and tries to live the role they have been assigned. Only if the competitors try their very best to beat their real-life company will they create the necessary pressure that will drive their peers in the company team to perform at their best. Switching roles and viewing the company from the perspective of competitors is itself enlightening and helps many long-term managers take a fresh look at what their company is good at and where it can improve. In this sense, wargaming is also a means to overcome organizational blindness. The second important aspect of successful execution is to stick with the schedule and agreed deliverables. There is nothing more annoying than a situation, in which teams are on time and deliver the agreed level of quality while one team lags behind. If this happens, the entire wargame is in danger of running out of sync, which will jeopardize the objectives.

One of the most critical aspects of successful wargaming is to capture what has actually been seen during the game and turn it into actionable measures. To this end, all available sources should be evaluated, such as the mail log, which chronologically captures all interactions between teams. Another source of information is the various plenary presentations by the teams. Here the actual presentations can be used, or as we have done in several cases, the presentations can be videotaped for later review. In any case, if coaches have been used, their input should go into the evaluations as well, because they oftentimes have great insights into team dynamics, the types of discussions that went on during the breakout sessions, and the decision-making process in the assigned teams. Yet another source is the qualitative evaluations by the market team and the quantitative results calculated by the control team. At the end of the actual game, time should be invested in a thorough feedback session or so-called hot wash up to capture the fresh impressions of participants and coaches as well. After all, participants have left the site of the wargame, the moderating team, which has administered the game, should invest a substantial amount of time to go through all the source material, and systematically comes up with insights and recommendations. The results are then usually put together in a detailed report that can be distributed among the participants of the wargame. In a final step, and based on the consequences identified, ideally in collaboration with the relevant people from the company who are responsible for the respective areas, specific measures and tasks have to be generated, which then need to be assembled into an implementation plan with an assigned person responsible for the implementation and set implementation date. Once set up, the implementation plan can then be monitored and controlled on a regular basis until the changes have all been made.

4 What Is the Significance of *Business Wargaming* Today?

Over the last decade, the use of business wargaming as a method to improve decision-making in the business environment has increased, and more companies are making use of this methodology. While originally used predominantly for questions revolving around market entry or competitive strategy, now questions relating to the reorientation or fundamental redesign of business models have been added to the list of applications. In our view, that is a direct effect of the increase in uncertainty to levels where traditional decision-making and planning methods no longer can do the job. Why? Because previously fundamental assumptions, such as which industry am I in, were never challenged before like today. After all, 7 years ago, hardly anyone would have thought that a previously unknown company called UBER would shake up the heavily insulated and regulated cab market, or that AirBnB would become the largest provider of guest rooms without owning any buildings. Therefore, if all your business tools were geared towards optimizing your operations in a particular industry, they might not be obsolete but certainly cannot help you if the industry they were originally designed for is fundamentally changing. A layman's view could therefore be: "Well if my tools, don't work anymore, I might as well try something new and play out what happens!"

However, the business of wargaming itself is changing as well! Yes, the design principles should still be true, and the wargame will still be set up to answer specific real-world questions, but the way in which wargames will be prepared, executed, and evaluated is about to undergo significant changes as well. So what could these changes look like?

5 The Future of "Wargaming"

Technology is and increasingly will be affecting the way wargaming will be prepared, run, experienced, and its findings transferred back into the real world. When looking at the possibilities, we subsume a broad range of serious games, simulations, and different types of wargames with varying degrees of sophistication under the general term "wargaming." While doing so, we recognize that the application of technological developments will affect the different types of wargaming differently. While serious games can greatly benefit from an increase in liveliness (i.e., through VR) that will allow the player to immerse in a virtual world so realistic that it is hard to discern from real life, business wargames, explained earlier in this chapter, may not benefit to the same degree from that same technology. However, they might benefit greatly from completer and more realistic "data worlds" upon which to base decisions. In any case, the future of wargaming will be exciting. We are currently seeing a mixing of multiple industries: wargames, simulations, and gaming. We are also seeing new technologies such as AI, cheap and widely owned computing devices, and virtual reality. From the new developments in these industries and new technologies, we can expect to see significant improvements to wargaming applications in business.

Specifically, wargaming will get better in three key areas: liveness, accessibility, and transferability.

Liveness refers to how immersive and responsive the wargaming experience is. Imagine if your wargame allowed you to chat with and see how your customers change over time. Accessibility refers to how these technologies make it cheaper to create and play wargames so more people can access them. Imagine if every employee could play a tailored wargame to their job role instead of just providing wargaming to executives. Transferability refers to how lessons learned from wargames can be applied to real-world environments. Imagine if the wargame never stopped and the wargame would continue to simulate, analyze, and give feedback about your decisions in the real world. These changes are already happening today with companies building personalized simulations for employees, business-specific games, and algorithms to analyze business decisions. For each of these three areas, we will look at how they will change wargaming in the short and long term. We can see a general progression as AI and software start out as just tools for humans to use in wargaming. Then, they will gradually evolve into autonomous assistants and coworkers that can cooperate in the wargaming process with us.

Starting with liveness, wargames will feel like playing a video game in the future with better tools, interactivity, and real-world data integrations. Forio is a company that builds software tools to let anyone create interactive business simulations to give players a sense that they are immersed in the experience and get to test their decisions and see the results of their actions. Similarly, Adventr is building tools to make it easy for anyone to build and share interactive videos. Now, wargame designers can better incorporate video content into their simulations. We can also expect to see AI as a counterparty and judge in wargames. Large language models like ChatGPT provide a realistic and rational AI player that can compete against human teams and also automate the judging process. In the future, we will see better integration of real-world data and metrics into wargames. Imagine if wargames could include news headlines from the real world and if an AI could go out and gather up-to-date data from within the company while the game is happening in real time.

With AI and software continuing to improve, wargaming will become cheaper and more accessible. In the past, it took a team of analysts up to 3 months to assemble, analyze, and structure relevant information until it could be condensed into a playable "game book." This process could be significantly expedited with technology. As a result, the process could take only days or weeks using new software tools. Software and AI will help us automate the busy work of creating a wargame such as sourcing data, creating scenarios, and developing materials. This means wargames can be more accessible to everyone including employees as they become cheaper to run. Currently, STRIVR is a company building VR simulations to train employees in companies like Walmart and Verizon. For example, they created a training experience to teach store employees how to safely respond to an active shooter situation. GoPractice is another company that built a simulator to teach product managers how to make data-driven decisions. Users can make decisions based on real-world data and see the results of their actions happen. Wargames and training simulations like these will become accessible to every business and industry. Imagine if AI could create tailored wargames for anyone in the company without needing additional programmers to predefine and design scenarios.

Lastly, it will be easier to transfer wargame learnings to the real world. Today, wargaming scenarios do not fully capture all the nuances of the real world, and they do not close the feedback loop. Today, wargames do not consider the real outcomes that come from implementing learnings or strategies from running the wargaming process. Imagine if the wargame never ended so that businesses could rely on a wargame 24/7 to simulate, strategize, and reflect on their next decisions. Currently, we see data-driven efforts to simulate logistics by companies like Flexport which run live simulations to optimize logistics decision-making. Providing live data is immensely valuable since wargames traditionally take months to prepare for and the data prepared can be stale by the time the wargame is run. AI and software can help us better collect real-life data and keep data fresh to be used in wargames so that the scenarios better reflect reality. Software tools can also make it possible to get real data during the wargaming process. For example, companies like Apify build easy-to-use scrapers using ML to pull live and real data from multiple live sources. Companies like Pollfish make it fast and easy to get survey responses from real people. However, there are limitations to the use of real-time data today. In wargames, some of the data (i.e., overall economic development based on different scenarios, key development, or shocks such as a war or a pandemic having an impact on delivery times, oil prices, free flow of the workforce, etc.) need to be simulated or extrapolated. In the future, the computer could play out an entire scenario based on a number of input variables and data points to simulate how they would change based on user decisions. In doing so, the computer would provide game participants with a very comprehensive and reasonable sense of having quantitative data. This task is traditionally performed by (human) members of the control team, who have limited bandwidth to update all parameters and compute extrapolated data. Here technology could compute a broader range of data and thus simulate an environment, where one of the key skills of the participants is to discern relevant from irrelevant information, just like in their everyday job.

Going forward, technology will also make it easier to capture learnings and insights and analyze them during a war game. This could go as far as analyzing intra-team dynamics as well as inter-team dynamics, strategic decision patterns, communication approaches, or all kinds of sensitivities related to any set of parameters to be investigated.

In the long term, we can expect wargames to automate the human busy work out of the equation and drastically improve the process for everyone.

6 In Which Areas Should Further Research Be Conducted?

Further research could help us to better understand the success factors of *business wargaming*. This could be done by means of a structured study of the experiences of companies as well as the military. The study could be built around the initial variables: quality of the game design, selection and background of the participants, quality of the moderation and coaching, application of technology towards liveliness, accessibility, and transferability, as well as the quality of the control models and the follow-up process. Here it would be particularly useful to explore the efficiency frontier between effort and return so that resources for the preparation and execution of *business wargaming* are bundled where they bring the most value. In view of the rather high financial and time expenditure for good *business wargaming*, this is quite justified.

A further field of research could be to investigate how the methodology can be used better with regard to the desired cooperation between universities, public institutions, and companies. Here, one should look at where the barriers to the application currently lie; why the approach is used more often in the Unites States vs., for example, in Europe; and how a possible, if perhaps only subjective, conflict of goals between the companies and the public institutions can be overcome. Optimal forms for organizing the implementation and sharing the costs would also be examined.

Finally, yet importantly, an effort should be undertaken to better understand and explore how emerging technologies, such as Artificial Intelligence, could be applied to wargaming. We have provided some initial ideas and examples, but a systematic study of potential applications of AI and other technologies would be most helpful. If a professionally prepared and conducted *business wargaming* previously cost anywhere between USD $250,000 and $750,000 (including research cost and manpower), this number could be significantly reduced by applying technology going forward. With a lower financial barrier to entry, the wargaming methodology should gain further momentum and become a standard tool for exploring complex and uncharted territories at different levels of any organization.

Reference

McGonigal, J. (2011). Reality is broken: Why games make us better and how they can change the world. *Jonathan Cape*.

Daniel F. Oriesek is a Swiss general staff officer, civil servant, and entrepreneur. After many years in banking and strategy consulting, while pursuing his military career as a reserve officer, he served two tours in the Balkans and in 2014 became a full-time employee of the Swiss Department of Defense, where he served several roles at the tactical, operational, and strategic level.

David Song is a software engineer creating artificial intelligence (AI) software products in San Francisco. He currently works on early-stage AI with Elad Gil and previously worked at Mem.ai. He studied computer science at Stanford University, specializing in AI and machine learning.

Teaching Imagination and Future-Shaping Skills: What Do Universities Offer Students to Help Them Imagine and Invent?

Ted Selker

Abstract This short chapter reviews some of the ways that imagination is promoted and sometimes learned in and around schools. Parents and teachers try to help children view, review, and work with things that expand their purview. At school, we are shown places, things, and techniques we never imagined before. Such envisioning can be done in ways that are prescriptive and limiting to the course's test on the topic. Today many are pushing to make the experiences more explorational by exposing students to possibilities, places, and things, and techniques can be used with material already known and motivation to find out more.

Imagination is visualizing things that do not but might or might not be possible. It is a critical part of creativity. Some worry that people are only creative when they are children. Some worry that only some people are creative. There is a rich literature around training people to think outside the box. If you listen too carefully to Sternberg's tomb on creativity, however, you will come to believe that creativity cannot easily be taught and is difficult to transfer across domains.

Actually, we all are creative in some areas. Somehow everyone can dance, be it building a shelter, the way you pack, the way you pick which shirt to wear, the way you fix the closure of a bag, everyone makes some creative decisions every day.

Outside classes, schools can offer mentoring experiences where students work with projects ranging from research or planning to constructive. Again, these experiences can be prescriptive such as asking a person processing a circuit board to dip the board in a solution. It can be exploratory, asking a person to learn how this equipment works so we can etch a circuit board.

More and more schools require some sort of internship for students, giving them as many chances to see how professionals do the work they are hoping to learn to do. Antioch specifically was known for being all about internships. Again, the internship can be prescriptive furthering the idea that you have to use procedures or it can be expanding, helping students discover new things in their projects.

Graduate programs again can be prescriptive: "Take these courses to get mastery and receive a MS degree." They can alternatively be thesis oriented, giving students

T. Selker (✉)
Palo Alto, CA, USA

T. Carleton et al. (eds.), *Inventing the Almost Impossible*, Future of Business and Finance, https://doi.org/10.1007/978-3-031-36224-8_7

a feeling that the courses and the lab work are in service of discovering a path that you want to pursue around a project that is sized well for that journey.

Schools should and try to work to expand students' abilities to imagine in new ways. They can include courses in design process, envisioning, entrepreneurism, and interdisciplinary activities such as are practiced at the Hasso Plattner Institute of Design at Stanford University.

Imagine games are typical and encouraged for babies and children. It might seem trite, but practice is helpful in everything. We get used to imagining the same things, should think about ways of imagining differently.

1 Projects as School

Hampshire College and Evergreen State College were both set up based on a book that showed that the curricula of Amherst College stifled the most creative students who wanted to take the initiative. The students are required to do a small project of their choosing approved by a committee of faculty in each of the college's five schools. This Division 1 project is then described and "defended" in front of a committee of faculty in the school it is designed to teach. They are then required to do two Division 2 projects that are much bigger in two of the schools. Finally, they get a degree when they complete and defend a Division 3 project, or thesis, in one area. The students at these colleges are encouraged to take classes only when they support the education. Other universities allow or require capstone projects for undergraduates too. These can be transformational experiences in helping a student develop their sense of goals, project, planning, and scoping.

2 Projects in Classes

A project-based class might include a project that a team works on together. Many times, the teachers start by teaching students the tools they will need; they then give the students the problem and describe several solutions. This focuses the students on the project-based issues of scoping, planning, coordinating with others, and presenting the idea. These are difficult and important skills to learn. Often these experiences are memorable for their bonding potential and intensity. In such a forum, the constraints work to focus the participants.

The same things that focus the participants can limit their learning of other things. The variability in engagement that occurs in many project-based learning activities led me to modify my project-based teaching approach. In a computer science class, we want everyone to learn the basics of computer science. In a design class, we want everyone to learn design. In a project management class, we want everyone to learn project management. If in a project one person takes on the project management role, another the coding role, and another the design role, they are each learning very different skills of management, coding, and design. I also found that a strong or even flashy personality in a group would limit the imagination, creative contributions, or ownership of the project by others. In a very real way, they took different classes. They should get credit for things they learned. Typically, we either think we can gloss over the differences in what different people learn in a group project and say they learned from the others or have two classes that collaborate siloed by topic. I sought a different model.

The model for my Android product development course was to enable each student to own their creative solution. As a graduate level class, I told them, "You know your computer science and I will not be teaching syntax of Java in this class." The whole course used the same tools and had to use the same programming environment, create posters from a template, create a one-minute video, and show their work as a demo. The bonding experiences were the similarity of tools and products they could help each other with and the group discussions about their progress. My bonding tool for the group was to set them up so with one button click they could all download a complete program development system with a phone simulator for the development phones I handed out. By everyone having the same and functional tools the first day of class, they had that to share. Everyone was to be responsible for their own project. Everyone had to come up with five ideas in the first week. I have never faced as much blowback as asking them to come back with ideas. I received multiple paragraph emails stating that no teacher had ever required them to produce ideas by themselves. They complained that to teach these skills I needed to give them programs to write that emphasized the skills.

Instead, I lectured the first day about all the apps that existed that used a special feature of the phone: the phone has so many sensors that are not found on a computer, the phone has multiple radios not found in a computer, and the phone can be used anywhere. I gave many examples of novel scenarios for phones. They had to come back having researched alternative things they could work on having researched what has been done, what cannot be done easily, and what could be done with less than 80 h of programming.

The second class was only getting reactions to the ideas people brought. Is it really too much for the time you will have this semester? Is it really interesting relative to things other entrepreneurs and researchers are doing? What are the tradeoffs of your five ideas? Pick two ideas for which you will try to prototype the hardest part this week.

By engaging the whole class, people experience a range of architectural, usability, scenario, business, and project sizing issues more deeply.

During week 3, students came back having asked help from each other and me. Their eyes are more open to the difficulties in their visions and the realities of implementation. The discussion of so many alternatives gives the class an overview of the programming issues in mobile development, ways of making decisions about direction, and scoping.

In week 4, the students need to come back with a mostly working version of a project. This includes a demonstration that all the hard parts of the project are being tested. They discuss alternatives such as the fast Fourier transform analyzer that one student used to tune a guitar. The exposure to other students' projects helps people know who to go to for help on the rest of the app, how to debug their app, and how to simplify and re-scope their project. The assignment now is to start making a poster and a video. I lecture about communication approaches in which I talk about the need in a video or for a viewer of a poster to be tantalized repeatedly and at least every 20 s of viewing.

Week 5 is spent talking about the poster: title, images, references, and results to get them thinking about the story. We show and discuss what can be done in a 1–2-min video.

We talk about the many problem apps that are in need of rescue.

During week 6, we are showing working apps. We are talking about sketches for videos; we are showing sketches of posters.

Week 7 is when we present to others.

The point of the short summer intensive smartphone development example is to show several ways I got every student to individually learn deeply the important parts of the curricula: idea generation and pruning, pivoting, scholarship, development tools and practice, collaborative problem solving, and presentation.

As a student I took independent study classes to try my ideas with a professor. It is much lonelier than my smartphone class. I did not have others around using the same tools, asking similar questions, and debugging the problems I was or there to present to. I also rarely had the luxury of changing my direction.

Research projects are even more constraining; they have a predetermined hypothesis that was formed before the project started. This does not create imagination.

The spurt of fun happens when people bring me problems. At theMIT Media Lab, Master Card brought the need for a secure RFID. I designed several. The immediacy of need of a request can help the imagination.

In my Industrial Design Intelligence class, I held an idea exploration session. This lecture consists of me bringing a problem into the class and soliciting solutions. The thing I did was classify and ask for more. So when an idea about a car door that will not hurt a bicyclist is soft, I put it in one column; when another person adds that it might break off, I put it in the same column. I then ask, what other ideas can we generate. Someone says the door can slide back, another says under another says over the car. The non-hinged door is now a category column. I ask for more approaches. One person says, "Let's use a camera to find the bike and lock the door." Another adds, "Let's signal the bike that the person is opening the door." By adding different category columns for different approaches, I was trying to inspire people to imagine the maximal alternative categories of ideas. The goal was to celebrate and judge an expressed idea as part of a column or to create a new column. Each new idea pushed people to try to imagine a new approach. These sessions were generative and exciting. They were among the best teaching experiences I have had.

Imagination is not vague; it has a story.

Imagination is not isolated; it is context.

Imagination is not singular; it comes in relation to many possibilities.

The maker movement may have come out of the old world of books like The Boy Mechanic, projects in periodicals like Popular Science, and a whole genre of creation-based education exemplified by making toys like erector sets and LEGO. Erector sets, chemistry labs, and Legos dominated the childhoods of many post-war children teaching them about creating. As a child, I ogled the Heathkit catalog. It was filled with kits you could put together to create every electronic gadget on the market from the components. Most of these project-based toys did not have deep pedagogy (but aspired to). Project-based activities with computers require more concentration and planning than other project-based activities. If the programmer has not thought through their program, it does not work. The only way to make a program work was to imagine it, design it, build it, and debug it. Until every aspect of the idea was fleshed out and translated into the syntax of the computer, it did not work.

I like to think that the maker movement followed this ideal. You must understand the parts and their relationships to make something with a laser cutter or 3D printer. You must at least understand the interfaces and instructions of Sparkfun parts to make something from them. The maker movement may not have embraced invention, in that scholarship was not the point. It embraced creating a system from parts. The idea of makerism is that we can make almost anything with parts and software. The idea was for this to spur imagining more.

The missing ingredient in many "how to make anything" courses and maker shops is needs finding. Without a good understanding of what is needed, it is hard to have a vision of what should be. If there is one thing that I criticize most in most curricula on makerism, it is the way that scholarship and needs finding is done. Many curricula celebrate asking people what they want. This is great for learning a bit about an area you know little about. It is not great for becoming an expert. My whole life I have walked through buildings designed by people that did not know what it is like to use the buildings. Many famous buildings echo loudly, making people feel small and disoriented. How is it that some companies want every entrance and building to be more important and impressive than you should deserve? How is it that most hospitals are mazes of confusing corridors you typically traverse when already stressed and disoriented?

Anything we make should be guided by an appropriate vision for its use, not just the presentation of the project. This approach requires context and empathy. These ideals are different from clipboard lists recorded with experts. The experts remain the knowledge sources, but you must internalize by putting down the clipboard and being the expert to know what is needed.

In classes, we use role playing, scenario creation, props, mockups, and prototypes to get a feel for ideas. This is the stuff that feeds imagination. Often missing are the alternative roles, scenarios, props, mockups, and prototypes that can make us reflect on options as we finalize a solution.

One of my first successful project-based educational experiences was getting 30, 3-person teams to each make an AI-based computer programming help system that adapted to the user's demonstrated knowledge. I started the course by spending 6 weeks teaching many AI approaches to representation, reasoning, and learning. I told the students they had to use some of these ideas in their resulting system. I created and gave them a window-based UX structure to present the information about

LISP in. By the assignment constraining the context, the 30 groups all made systems that made sense to each other. They all made different kinds of solutions that helped them all see different possibilities. The various groups collectively built a landscape of alternatives that became more interesting than any one system could have been. The resulting lessons from the alternatives were strong: an AI learning system that responds immediately is more engaging than one that might change things in the future. This example showed me the value of alternative ideas to help form a better understanding of the problem trying to be created.

My approach is to always be learning every tool in case it might be useful. When I find a problem I ask, "Do I have some things I know of that can help solve this? Can I put together techniques known to a person in the art to solve this?" Then I ask, "What if I had new techniques, how would that improve the solutions I can create?" That is the moment of serious imagination; "What are the futures that allow me to build better solutions?"

Tamara Carleton and Bill Cockayne's Moonshots courses and teaching enable students to focus on modern horizon thinking, taking imagination in a different direction. The idea is to think about what could be done now, a bit in the future, at a time when a novel invention would be needed to do it and at a time when most of today's constraints are no longer constraints. This takes imagination in a very different landscape of expanding scopes that stretch people out of their what they know can be done cocoons.

Still, we go back to literature about creativity and productivity. Mihaly Csikszentmihalyi taught us that we should work on things at a level in which we know enough about what we are doing and how to do it to make progress. He calls this state Flow. Jeff Dow did a series of demonstrations that show that considering multiple solutions in parallel outperforms interactive improvement in design. Sternberg wrote a whole handbook on creativity that covers many experiments that show that people can improve any skill they work at in a specific domain. He showed much worse results in teaching people to be creative across domains. I find the concern that people do not transfer creative skills seems cynical. II see encouraging people often spurs them to new successes.

Universities are places where we can learn anything. A catalog of courses serves hundreds of different degree alternatives. The catalogs of universities spurred my imagination as I began my university studies. Then the reality comes through: an ABET approved engineering degree has so many required courses that a student's schedule is full of physics for engineers, statics, systems, and other classes that do not sound so imaginative. Business students take none of these but are busy learning how to do accounting, marketing, sales, and so forth. So where is the imagination in such a university? As well as the Antioch and Hampshire colleges described above, Stanford's Hasso Plattner Institute was another answer to this, bringing people across curricula for exercises in problem solving. This is great at least at getting people to see that there are other ways of thinking about things. I dream that we are starting to move farther.

The space between engineering and business curricula is technology invention. When we understand and get better at teaching the skills of entrepreneurship and technical evaluation of new ideas, we will have a special class of skilled workers. These workers will be able to evaluate engineering solutions for the many steps needed to move them from a prototype to something that can be sold, built, supported, and improved to make a continuing and productive enterprise.

We naturally enjoy and support imaginative exploring play in children. Learning to play a piano is not about imagination; it is about skills, craft, and mastery. Learning to dance specific ways is also craft and mastery. These crafts do require envisioning and concentration. What they do not easily support is the imaginative expansive self-guided creative exploration. Interpretive dance or creating new dances requires imagination. We naturally gravitate to focusing on being thorough and dependable in learning the trades in our young adulthood. Lest we lose the ability to form novel problems and solutions we must work to keep parts of the imaginative exploration in our schedule always.

Recommended Reading

Burleson, W., & Selker, T. (Eds.). (2002). Creativity and interface. *Communications of the ACM (Special Issue), 45*(10), 88–90.

Carleton, T., & Cockayne, W. (2023). *Building moonshots: 50+ ways to turn radical ideas into reality*. John Wiley & Sons.

Carleton, T., Cockayne, W., & Tahvanainen, A. (2013). Playbook for strategic foresight and innovation.

Kaufman, J., & Sternberg, R. (Eds.). (2019). The Cambridge handbook of creativity (2nd ed., Cambridge handbooks in psychology). Cambridge University Press.

McKim, R. H. (1972). *Experiences in visual thinking*. Brooks/Cole Publishing Company.

Resnick, M., Myers, B., Nakakoji, K., Shneiderman, B., Pausch, R., & Eisenberg, M. (2005). *Design principles for tools to support creative thinking*. Report of Workshop on Creativity Support Tools. No. 20.

Selker, T. (2005). Fostering motivation and creativity for computer users. *International Journal of Human-Computer Studies, 63*(4–5), 410–421.

Selker, T. (2008). Fostering motivation and creativity for computer users. In *Proceedings of NSF CreativeIT Workshop at Arizona State University*.

Shneiderman, B., Pausch, R., Selker, T., & Eisenberg, M. (2005). *Creativity support tools: A workshop sponsored by the NSF* (pp. 25–36).

Ted Selker is an American computer scientist known for his user interface inventions. Ted spent 10 years as an Associate Professor at the MIT Media Laboratory where he created the Context Aware Computing group, co-directed the Caltech/MIT Voting Technology Project, and directed the CI/DI kitchen of the future/design of the future project. Ted's work strives to demonstrate considerate technology, in which people's intentions are recognized and respected. He is well known as a creator and tester of new scenarios for working with computing systems. His design practice includes consulting to help dozens of startups and large companies, speaking engagements, and innovation workshops. His successes at targeted product creation and enhancement led to his role as IBM Fellow and director of User Systems Ergonomics Research at IBM. He has served as a consulting professor at Stanford University and taught at Hampshire College, University of Massachusetts at Amherst, and Brown University. He worked at Xerox PARC and Atari Research Labs.

Future Labs: Making the Future Tangible Today

Andrew Paice and Elena Malakhatka

Abstract This chapter summarizes the important knowledge of open innovation (OI), Design Thinking (DT), and living labs (LL). It highlights the need for collaboration and co-creation among stakeholders from different backgrounds, perspectives, and experiences to co-generate new business opportunities. The role of users in the innovation process is emphasized, with the concept of user-driven innovation and participatory design being crucial. The authors also introduce two case studies, iHomeLab and KTH Live-in-Lab, as examples of living labs and field tests that enable testing of innovative technologies and methods in real environments. The feasibility of innovation in human, technical, and economic domains is highlighted, and the potential of technology, data security, ethics, and new processes for enhancing innovation capabilities is mentioned. Overall, the chapter sets the stage for discussing the importance of openness, collaboration, and user-centric approaches in the future of innovation.

Many theories explain the nature of innovation and point to practical applications. This topic is broad and can be considered from different angles and at different levels. However, the percentage of innovative ideas implemented in life is still relatively low. For example, only one in 3000 innovative ideas in the ICT sector reaches the market (EU Commission). This means that considerable efforts to create something new remain unrealized. The reasons for such low rates can be interpreted in different ways. One of the reasons that the academic community has been discussing quite seriously lately is the gap between research laboratories and the adaptation of acquired knowledge to the canvas of real life. Some researchers call this phenomenon a Pre-commercial Gap (Macdonald, 2004) or Chasm (Moore & McKenna,

A. Paice (✉)
iHomeLab, Department of Engineering and Architecture, Lucerne University of Applied Sciences and Arts, Horw, Switzerland
e-mail: andrew.paice@hslu.ch

E. Malakhatka
Department of Architecture and Civil Engineering, Division of Building Technology, Chalmers University of Technology, Gothenburg, Sweden
e-mail: elenamal@chalmers.se

T. Carleton et al. (eds.), *Inventing the Almost Impossible*, Future of Business and Finance, https://doi.org/10.1007/978-3-031-36224-8_8

1999). Today, the academic community is aware of the need to bridge fundamental research and real life and create new forms and formats of research laboratories, such as living labs or innovative laboratories (iLabs). According to (Halila, 2007), to realize an innovative idea in real life, three things are required: a process of integrating new technology into the existing context, a network of actors that connects users of this technology in a new way, and consequentially a new social, economic, and environmental reality shaped by the adoption of the new technology.

Nowadays, we are entering an era that will be defined by the digitization and connection of everything and everyone. This global trend will lead to the emergence of a new global-local socioeconomic order in which digital products and services are offered globally, adapted, and delivered locally. We believe that such new orders should be collectively co-created with a certain degree of transparency, inclusivity, and openness. In these kinds of processes, openness is crucial for the innovation process. It is vital to gather various points of view, which can lead to the successful development and implementation of new ideas and generate new and unexpected business openings. However, to be able to collaborate and share in a multi-stakeholder environment, different levels of openness between stakeholders seem to be a requirement. Eriksson et al. (2005) propose open collaboration between people from different backgrounds, perspectives, knowledge, and experiences to stimulate creativity and create new ideas that can be applied and benefited through use. More people, including consumers, need to be involved in the innovation process. This is argued by Von Hippel and Katz (2002), who stated that users are often the source of innovation. The concept of user-driven innovation (Urban & Von Hippel, 1988) suggests that users are capable innovators. Thus, it can be argued that the participation of end-users or consumers in the innovation process is essential. They should be a vital part of the innovation system.

Participatory and user-centric design ideas are not new, but the level of complexity with which today's developers of future laboratories are confronted demonstrates a qualitatively new level. Over the past decade, industrial innovation labs at public universities, the living labs, and a New European Bauhaus movement have gained momentum. These new formats of future labs testify to the modernization of laboratory infrastructure and the emergence of a new mindset and a culture of innovation where openness, collaborativeness, and democratization are becoming the new normal.

1 Open Innovation

In most research and innovation laboratories, multi-stakeholder platform strategies are becoming one of the key priorities. In the EU's Innovation Union agenda, they are regarded as a crucial part of the common research agenda Horizon 2020. Academia and science are asked to be involved in processes of open knowledge creation, open science, and open innovation (Chesbrough, 2003) referring to different fields like industrial leadership, societal challenges, and science. With an increasing awareness that the traditional model of innovation is becoming obsolete, a new paradigm of "open innovation" (OI) has emerged, connecting internal and external sources of information-rich environments (Chesbrough, 2003). Nowadays, open innovation (OI) plays a crucial role in improving organizational construction and strategic maintenance as well as enhancing the competitive advantage of firms.

The original definition of OI stressed that "valuable ideas can come from inside or outside the company and can go to market from inside or outside the company as well. This approach places external ideas and external paths to market on the same level of importance as that reserved for internal ideas and paths" (Chesbrough, 2003). Both innovation scholars and Chesbrough have modified its original definition, with the latter stressing the intentionality of knowledge inflows and outflows: "Open innovation is the use of purposive inflows and outflows of knowledge to accelerate internal innovation and expand the markets for external use of innovation, respectively" (Chesbrough, 2006).

Open innovation is based on the idea that, within the modern competitive context in which firms have to operate, the linear model of innovation (Urban & Von Hippel, 1988) is no longer able to be applied to the innovation process. Today's organizations have to collaborate with external stakeholders through the iterative exchange of knowledge, technology, and resources across their boundaries (Galati, 2015; Galati & Bigliardi, 2017). In other words, to stay relevant on the market, a single organization cannot innovate in isolation, it has to engage with different types of partners, ranging from companies with the similar visions to customers, universities, research centers, and even competitors (Bigliardi & Galati, 2017), in order to acquire ideas and resources from the external environment (Laursen & Salter, 2006). Open innovation promises to increase internal innovations through the involvement of external parties and the commercialization of new ideas, which can create new value for business. Value creation for the industry is by the commercialization of the latest technology and patents (Chesbrough et al., 2006).

One of the most important transformations, which OI can facilitate, is the conversion of rigid organizational boundaries into semi-permeable ones which enable innovations and knowledge to move freely between the organization and the external environment. These inflows and outflows of knowledge do not spontaneously occur; organizations must generate opportunities and mechanisms for identifying existing knowledge and transferring it into or outside the organizational boundaries (Spithoven et al., 2010).

It is not an easy task to define the degree of the openness of a specific organization. Scholars propose two variables are considered that represent the degree of openness for an organization:

1. The number and type of partners with which the company collaborates.
2. The number and type of phases of the innovation process that the company opens to external contributions.

By crossing these two variables, four basic modes of open innovation are identified: closed innovators, open innovators, specialized collaborators and integrated collaborators (Lazzarotti & Manzini, 2009). Most of the research and innovation labs affiliated to different universities have a higher degree of openness compared to the industrial R&D labs. There are many reasons to explain this phenomenon, one of each is the more open mindset and less competitive culture in the academic community. Creating a culture that values outside competence and know-how is crucial for open innovation practice. The culture of open innovation is built on values such as curiosity, creativity, flexibility, and diversity, because the open dimension requires values such as openness, trust, responsibility, authenticity, and sustainability (Matricano, 2018; Phillips, 2010). Innovation culture has five characteristics: (1) vision, (2) network of knowledge, (3) inspiration and leadership, (4) Freiraeume (which means room for ideas), and (5) creativity and risk-taking (Schumpeter & Nichol, 1934). By this we can conclude that open innovation is a necessary attribute of modern research and innovation labs, which can bring an inclusivity and diversity into the process of new ideas formation. We also agree with Enkel et al. (2009) that the future of innovation processes lies in an appropriate balance between open and closed innovation approaches, because too much openness can lead to a negative impact on companies' long-term innovation success, loss of control, and loss of core competencies, whereas a too closed innovation approach does not serve the demands of increasingly shorter innovation cycles and reduced time-to-market.

Another area where OI can contribute is in dealing with disruptive innovation. The seminal book Christensen 2009 outlined the process by which innovation may disrupt an industry. Taking case studies from fast-paced industries (Hard Disk Manufacturing) and slow-paced industries (Mechanical Excavation), he defines disruptive innovation as being an innovation which leads to a radical shift in the market, where typically most successful incumbents are replaced by new market entrants. The disruption is also characterized by the innovation starting in a new or niche market and then disrupting the larger market when performance and cost becomes competitive. This leads to the innovator's dilemma: the technology which can put you out of business is most likely to be less competitive in your current market, and investments in this innovation decrease your business performance. The question is then how businesses can successfully explore new technologies and markets so as to secure long-term competitiveness in a disrupted market. OI provides a partial answer. Market leaders can investigate new technologies and markets via OI and integrate when they believe the technology is becoming successful. Another part of the answer is provided by living labs. They provide a mechanism to test such innovations and niche markets. A living lab is a prototype market of sufficient complexity to provide real data on how a technology can develop in a co-creation process with consumers and to provide the potential for emergence of unexpected uses of the technology and therefore also support the emergence of new business opportunities.

2 Open Innovation and Living Labs

New paradigms, such as open innovation (Chesbrough, 2003) and living labs operating as a User-Centered Open Innovation Ecosystem (Pallot et al., 2010), bring into the conversation the role of users in the research and innovation process (Peltomaki, 2009). As was mentioned above, OI builds on a value co-creation process with users, and the result is expected to better solve customers' needs. Therefore, users are innovators, co-designers, co-producers, and entrepreneurs regarding new products and services (Pascu & van Lieshout, 2009).

Ballon and colleagues (2005) found that Test and Experimentation Platforms (TEPs) constitute a lot of opportunities for OI and co-creation with the users through participatory design. They identified six types of TEPs, namely prototyping platforms, testbeds, field trials, living labs, market pilots, and societal pilots (Pascu & van Lieshout, 2009). They also gave the following definition for a Living Lab: "An experimentation environment in which technology is given shape in real life contexts and in which (end) users are considered 'co-producers'." Living labs allow firms to involve users in the development of new products, services, or applications in a process of co-creation, because the average user, equipped with the proper tools, is the most suitable candidate to design a product or service (Leminen et al., 2012). Therefore, living labs offer an R&D methodology through which innovations are created and validated in collaborative real-world environments (Eriksson et al., 2006). Living labs bring experimentation out of companies' R&D departments to real-life environments with the participation and co-creation of users, partners, and other parties. Therefore, living lab projects are a specific case of open innovation where companies open up their innovation processes to users or customers (Schuurman et al., 2013), which can be linked to the user innovation paradigm (Von Hippel, 1976; Schuurman et al., 2013). Today, 212 living labs are members of the European Network of living labs (ENoLL). They are geographically located within the enlarged European Union and in other regions such as South Africa, Asia, and South America. All of them have the goal to involve users at the earlier stage of the R&D process not only as observed subjects but rather as a participative force for co-creating value.

Prior studies propose stakeholders as "providers" including educational institutes, universities, researchers, developers, or consultants bringing knowledge and promoting solutions for problems; "users" including end users, customers, or citizens to be studied or involved in innovation activities; and "utilizers" including a company or another organization utilizing achieved results; and "enablers" including financiers or area/city development organizations enabling innovation activities in living labs (Leminen et al., 2012). Furthermore, living lab can be viewed as "an arena for innovation. It is a structure and a long-term societal resource rather than related to a certain project. Within this structural framework, experiences, routines and conditions are built to develop ideas into innovations" (Arvidsson & Mannervik, 2009).

One of the important criteria for the living lab approach is that innovation activities should be carried out in a realistic, natural, real-life setting. Orchestrating realistic use situations and user behavior is seen as one way to generate results that are valid for real markets in living lab operations (CoreLabs, 2007). Relating realism to Checkland's real-world concept (Checkland, 1999) means that the "real-world" situation reflects people's interpretation of their current situation. People's interpretations and how they perceive the situation is related to people's worldview, or what they view as important for them; hence, what is viewed as the reality for one person does not necessarily mean the same for another person. This means that what is important and motivating for one partner is not necessarily important to another partner, which is a rationale for why it is crucial to involve a diversity of perspectives in the innovation process. To facilitate these types of processes, the Design Thinking methods are widely applied in the living labs.

3 Open Innovation and Design Thinking

A former President of the Design Management Institute Lookwood, suggests Design Thinking is "a human-centered innovation process that emphasizes observation, collaboration, fast learning, visualization of ideas, rapid concept prototyping and concurrent business analysis." Mintrom and Lieutjens (2016), whose emphasis is on the targeting market, assert "Design thinkers exhibit curiosity and empathy in their efforts to interpret how target populations engage with their world. They deploy various investigative techniques that have the potential to illuminate problems in new ways and indicate effective client focused solutions." Similar to a marketing approach, Design Thinking foregrounds the wants and needs of consumers, but Curedale (2013) notes that Design Thinking has moved far from being merely a tool in the marketing armory to designing things people actually want. These two definitions are connected with one core principle that Design Thinking is invariably user-centered and founded on some actionable insight. It relies on customer observation and uses visualization as a tool for communication. DT enables the creation of rich ethnographic portraits of customer behavior and trying to identify themes and patterns (unmet or under-served needs) from the observations.

One of the main advantages of the DT process is the ability to move through an innovative project to other areas of the organization based on the skillset of the designers. This kind of openness and flexibility of the method allows you to work with initially insoluble problems and, at the same time, achieve concrete and tangible solutions. This ability can enable collaboration in transforming organizations that want to move from a repetitive, mechanistic model to develop ideas, products, and services more in line with the speed of scientific and technological change (Lindberg et al., 2010).

4 Case Study: iHomeLab

4.1 General Introduction

The iHomeLab is the Swiss research center for building intelligence. It is part of the Department of Engineering and Architecture at the Lucerne University of Applied Sciences and Arts. The iHomeLab team conducts interdisciplinary applied research and consists of 25 computer scientists and electrical engineers, physicists, and mathematicians who research the use of the latest technologies such as Wireless Sensor Networks, Internet of Things, Ambient Intelligence, and Machine Learning to provide energy flexibility, security, and data protection in living spaces. The application-oriented research projects are carried out in close cooperation with partners from industry and business, co-financed by grants, and presented to the public in the iHomeLab Visitor Center. The iHomeLab has been registered with ENoLL as a Living Lab since 2015.

4.2 Reflection on OI and DT in the Context of IHL

With its focus on the interactions between occupants and intelligent systems integrated in their buildings, the DT approach has been fully integrated in the research methodology we apply. In particular, in the focus research area Ambient-Assisted Living (AAL), conducted in over 30 research projects since 2008, this approach has proven to be instrumental in conducting successful projects. The focus of AAL is to use technology to enable the elderly to lead an independent life in their own home. A typical challenge is to ensure that the needs of the solution fit the needs and the situation of the elderly, which have proved to be highly local due to their cultural, language, and demographic dependency. The fact that the elderly are partly technology averse or have problems assimilating new solutions complicates the problem. DT is an essential tool for introducing AAL innovations. It allows researchers to meet the elderly in their context and so better evaluate their needs and possible synergies or side effects of the new solution. We use DT to develop solutions, test them in field tests, and show them in the iHL Visitor center—providing dissemination to the public and potential partners.

Through our applied research projects, the iHomeLab operates as part of an open innovation network, taking the roles of technology provider, scientific investigator, and proof of concept designer. This approach—combining OI with DT—significantly increases the effectiveness of the research projects; however, overall market development is subpar due to its fragmentation, the needs of the elderly being very specific to local demography, society, and culture.

Increasingly, it is obvious that the introduction of a solution is dependent on the individual, their social network (relatives, friends, neighbors), support mechanisms (clubs, social networks, volunteer and professional services), and the local political situation. This means that the suitability of a solution can only be tested in a Field Test/Living Lab. For this reason, the iHomeLab joined ENoLL and applied living lab methodologies for the research projects and dissemination of the results.

4.3 Preliminary Results, Improvement Potential, and Future Visions

The impact of these ideas has been seen in several projects over the last decade. As an example of combining DT and field tests, we refer to the project IWalkActive, which won the AAL award in 2013. This project targeted the mobility of the elderly in outdoor settings. User Input was elicited at the beginning of the project and confirmed in the field trials. The project Relaxed Care showed the need to move a step further than the lab or living lab environment and perform long-term trials in the users' home environment. The typical disadvantage of a field trial to a living lab is the lack of measurement information. As IoT specialists, the iHomeLab is able to compensate for this lack by introducing additional sensors in the home, or as wearables. With this additional information, we can gather sufficient data to put together a complete picture of the user context. Relaxed Care provided a first step. In some projects, such as with Ella4Life or Living Well with Anne, which used an Avatar with a speech interface, the interaction and data collection could be integrated directly in the new technology. In others, such as the project Home4Dem, additional sensors were able to identify the Activities of Daily Living, which in turn maybe processed to allow conclusions regarding the state of mental health of the occupants. Combinations of IoT technologies—environmental and wearable sensors coupled with strict data security and machine learning—enables collection of quantified data regarding the interaction and reaction of the users to the new technologies. This is usually complemented by qualitative data won through interviews. The recent project RESTART has also introduced ChatBot technologies which conduct regular structured interactions with the subjects in order to get an up-to-the-minute picture of the user's subjective experience and yields higher quality data than interviews alone.

Another important trend in the area of AAL is investigating the impact of new solutions on the personal situation and social network in which the elder live, respectively, on how solutions must also be designed with not just the elderly, but also their network as key users. With projects such as Sam and Me or Kith'n'Kin, the role of technology in improving engagement and reducing isolation was investigated. In others, such as CabiNET, the focus was more on how to support the network and thus indirectly support the elderly. Such effects cannot be tested in the lab.

The current technical developments will provide more quantitative and qualitative data regarding the adoption and interaction of users with new technologies. Big Data technology will allow better interpretation and correlation of the data, which should lead to more effective DT and OI processes. We see augmented living labs and field tests as the key to unlocking more innovative processes in the assimilation of new technologies by enabling 360° qualitative and quantitative data collection in the natural environment.

5 Case Study: KTH Live-in-Lab

5.1 General Introduction

KTH Live-in Lab is a platform for accelerated innovation in the real-estate sector and for collaboration between academia and business in Stockholm, Sweden. Most test beds in KTH Live-in Lab are operated in real environments for testing and researching new technologies and new methods (Molinari et al., 2023). The purpose of KTH Live-in Lab is to reduce the lead times between test/research results and market introduction. In this way, KTH Live-in Lab aims to facilitate the advent of the sustainable and resource-effective buildings of the future. KTH Live-in Lab enables testing of products, services, and methods in real buildings, which results in a well-founded basis for changing structures and rules and increased use of new innovative technology. Tests in KTH Live-in Lab led to accelerated innovation.

KTH Live-in Lab encompasses a 300 sqm building permit-free innovation environment with alterable student apartments (Testbed KTH), which enables studies on the future's resource-efficient and sustainable student housing. The KTH Live-in Lab also receives property and user data from 305 common student flats owned by Einar Mattsson (Testbed EM) and from the KTH campus education building owned by Akademiska Hus (Testbed AH).

5.2 Reflection on OI and DT in the Context of KTH Live-in-Lab

The idea of an agile testbed for building-related Cleantech arose from discussions as to how we could eliminate the identified obstacles to increased innovation within the residential and construction sectors. KTH Live-in Lab is based on theory around Strategic Niche Management (SNM) and Multilevel Perspective (MLP) (Berkers & Geels, 2011: Schot & Geels, 2008). Both theories discuss innovation and technology shifts. They argue that players who are actively involved in the innovation process affect, through collaboration, the selection process of new technologies and the future trajectory of research and development. These theories emphasize the importance of demonstration projects, or testbeds, that provide partial shelter for new technological innovations, referred to as technological niches (Rip, 1992; Rip et al., 1995; Schot, 1992, 1998). Dynamic clusters thrive on the ability to test and verify products and services within protected environments known as technological niches

(Kemp et al., 1998). These niches facilitate increased interactions and knowledge transfer among various market participants, thus playing a pivotal role in the success of these clusters.

KTH Live-in-Lab can be used for testing and research in the building sector where inhabitants/users are engaged in product or service co-development and providing feedback to the innovating organizations. Our living lab provides open-innovation environments, which in combination with established open innovation ecosystems and respective stakeholder organizations can serve as an effective platform to foster the development and uptake of innovation in the building sector. We are focusing on the theoretical overlapping between two concepts of co-creation: co-creation as an innovation process as a part of open innovation theory and co-creation as a design process as a part of participatory design theory for the service concept development as a part of new service development theory. A few projects are specifically focusing on DT, such as "Sustainable Food System," where end-users co-created desirable food-related habits together with home appliances industrial designers and behavioral scientists. In a parallel project called "Kitchen," the users/inhabitants were involved in the process of redesigning KTH Live-in-Lab layout and together with the professional architects co-develop new co-living and co-learning spaces, which made it possible to rebuild KTH Live-in-Lab layout from 1.0 to 2.0. Another project is "Oura ring and sleeping comfort," aiming to improve the end-user sleeping quality, while operating the building in a more energy efficient way. This project included a series of participatory sessions with the end users. Currently, we are preparing a new big project called Sustainable Behavior Goals (SBGs) aiming to co-design sustainable everyday behavior narratives together with the users/inhabitants of KTH Live-in-Lab. This project will have a DT mindset as a core.

5.3 *Preliminary Results, Potential Improvement, and Future Visions*

We want to share a few preliminary results of using KTH Live-in-Lab as a testbed for innovation acceleration. First, the opportunity to engage multiple and diverse actors has enormous potential for cross-collaboration and new ideas formation. The process of open innovation brings a certain degree of transparency to the actors' network, creating a premise for sharing strategies, co-strategies, co-creation, and closer collaborations. Secondly, the practice of a bi-directional value exchange mechanism between different actors helps to identify both tangible and intangible values and enable better understanding of the relationships between actors, which can benefit the quality of relations and a more sustainable partnership in the future. Here we would also like to highlight that open innovation is a key driver of the diversified value proposition. Both OI and DT contribute significantly to shortening the time of innovation and bringing the concepts and ideas into the test phase immediately. Thus, the declared SNM and MLP methodology works not only in theory but also in practice and creates specific prerequisites for various types of innovations and improvements at all three levels: niche, regime and landscape (Geels & Schot, 2007).

We definitely see a lot of potential to improve OI and DT applications at KTH Live-in-Lab and experiment more with different user engagement models: from "expert's" mindset, where a user is seen as a subject, to more participatory mindset, where a user is considered as a partner. In addition, we would like to highlight the importance of actors' network analysis within a living lab lifespan and explore the innovation lifespan changing within time. Today's living lab process is happening in a more organic way, and a lot of processes happening as a "side effect" could be researched more.

We see the future of research laboratories in considering the laboratory as a service for research, testing, and facilitation of innovation process. The laboratory should become a center of attraction for various research needs: from student projects and start-ups to large companies and international collaborations. It is crucial that the laboratory becomes a kind of market space with various tools and can adapt to the needs of researchers and industrial partners. It is very important to be neutral and open to different actors. The diversity of minds attracts and creates an intellectual Brownian motion.

6 Conclusion

Innovation requires feasibility of the innovation in three domains: Human, Technical, and Economic. Human feasibility means that the innovation must provide a real benefit and be acceptable and usable and fit into the life context of the users. Technical feasibility includes the implementation of the technology, but also the existence of support networks—for power and information infrastructure, user support, and repair organizations. Finally, the solution must be economically sustainable—organizations must provide the necessary goods and services. In our opinion, living labs and field tests augmented by IoT are the only environment complex enough to test all these aspects. In the future, we need to bring this complexity forward in the innovation process—to enable with DT and OI—but without creating a situation so complex that innovation is stifled. New processes, insights, and methods, supported by technology and informed by considerations of ethics and data security, will help us gain insights and recognize potentially successful innovation early in the process. In this vision of the future, technology does not automate or replace DT and OI processes; rather it augments our capabilities to use them.

Today, the topic of democratization of innovation is relevant and essential for discussion and practical application. Concepts such as "privacy by design," "ethical design," and "co-creation"/"co-design" are becoming more and more popular, and most projects in these areas are successful. The ability to maintain a dialogue with both the private and public sectors might significantly increase the chances of technology acceptance, usability, and overall trust. Many companies today look to the future and actively create prototypes of future homes, future cities, and so on. We believe that the future should be co-created. Society should be involved in such a process more actively. Therefore, the Academic Community should play the role of a neutral partner in this complex but exciting process.

References

Arvidsson, N., & Mannervik, U. (2009). *The innovation platform: Enabling balance between growth and renewal*. VINNOVA.
Ballon, P., Pierson, J., & Delaere, S. (2005). Test and experimentation platforms for broadband innovation: Examining European practice. Available at SSRN 1331557.
Berkers, E., & Geels, F. W. (2011). System innovation through stepwise reconfiguration: The case of technological transitions in Dutch greenhouse horticulture (1930–1980). *Technology Analysis & Strategic Management, 23*(3), 227–247.
Bigliardi, B., & Galati, F. (2017). Family firms and collaborative innovation: Present debates and future research. *European Journal of Innovation Management*.
Checkland, P. (1999). Systems thinking. Rethinking management information systems. In *Rethinking: Management information systems: An interdisciplinary perspective* (pp. 44–56).
Chesbrough, H. W. (2003). *Open innovation: The new imperative for creating and profiting from technology*. Harvard Business Press.
Chesbrough, H. (2006). *Open business models: How to thrive in the new innovation landscape*. Harvard Business Press.
Chesbrough, H., Vanhaverbeke, W., & West, J. (Eds.). (2006). *Open innovation: Researching a new paradigm*. Oxford University Press on Demand.
CoreLabs, I. (2007). *Living labs roadmap 2007–2010: Recommendations on networked systems for open user-driven research, development and innovation, in open document*. Luleå University of Technology-Centre for Distance-spanning Technology.
Curedale, R. (2013). *Design thinking. Process and methods manual*. Design Community College Inc.
Enkel, E., Gassmann, O., & Chesbrough, H. (2009). Open R&D and open innovation: Exploring the phenomenon. *R&D Management, 39*(4), 311–316.
Eriksson, M., Niitamo, V. P., & Kulkki, S. (2005). *State-of-the-art in utilizing Living Labs approach to user-centric ICT innovation-a European approach*. Luleå University of Technology-Centre for Distance-spanning Technology.
Eriksson, M., Niitamo, V. P., Kulkki, S., & Hribernik, K. A. (2006). Living labs as a multi-contextual R&D methodology. In *In 2006 IEEE International Technology Management Conference (ICE)* (pp. 1–8). IEEE.
Galati, F. (2015). At what level is your organization managing knowledge? *Measuring Business Excellence, 19*(2), 57–70.
Galati, F., & Bigliardi, B. (2017). Does different NPD project's characteristics lead to the establishment of different NPD networks? A knowledge perspective. *Technology Analysis & Strategic Management, 29*(10), 1196–1209.
Geels, F. W., & Schot, J. (2007). Typology of sociotechnical transition pathways. *Research Policy, 36*(3), 399–417.
Halila, F. (2007). Networks as a means of supporting the adoption of organizational innovations in SMEs: the case of Environmental Management Systems (EMSs) based on ISO 14001. *Corporate Social Responsibility and Environmental Management, 14*(3), 167–181.

Kemp, R., Schot, J., & Hoogma, R. (1998). Regime shifts to sustainability through processes of niche formation: The approach of strategic niche management. *Technology Analysis & Strategic Management, 10*(2), 175–198.

Laursen, K., & Salter, A. (2006). Open for innovation: the role of openness in explaining innovation performance among UK manufacturing firms. *Strategic Management Journal, 27*(2), 131–150.

Lazzarotti, V., & Manzini, R. (2009). Different modes of open innovation: A theoretical framework and an empirical study. *International Journal of Innovation Management, 13*(04), 615–636.

Leminen, S., Westerlund, M., & Nyström, A. G. (2012). Living labs as open-innovation networks.

Lindberg, T., Noweski, C., & Meinel, C. (2010). Evolving discourses on design thinking: How design cognition inspires meta-disciplinary creative collaboration. *Technoetic Arts, 8*(1), 31–37.

Macdonald, S. (2004). When means become ends: Considering the impact of patent strategy on innovation. *Information Economics and Policy, 16*(1), 135–158.

Matricano, D. (2018). The state of the art of open innovation culture. *Exploring the Culture of Open Innovation*, 139–162.

Mintrom, M., & Luetjens, J. (2016). Design thinking in policymaking processes: Opportunities and challenges. *Australian Journal of Public Administration, 75*(3), 391–402.

Molinari, M., Vogel, J. A., Rolando, D., & Lundqvist, P. (2023). Using living labs to tackle innovation bottlenecks: The KTH Live-In Lab case study. *Applied Energy, 338*, 120877.

Moore, G. A., & McKenna, R. (1999). *Crossing the chasm, 1991*. HarperBusiness.

Pallot, M., Trousse, B., Senach, B., & Scapin, D. (2010, August). Living lab research landscape: From user centred design and user experience towards user cocreation. In *First European Summer School "Living Labs"*.

Pascu, C., & van Lieshout, M. (2009). User-led, citizen innovation at the interface of services. *Info, 11*(6), 82–96.

Peltomaki, A. (2009). Living Labs for user-driven open innovation.

Phillips, J. (2010). Open innovation typology. *International journal of Innovation science, 2*(4), 175–183.

Rip, A. (1992). Science and technology as dancing partners. In *Technological development and science in the industrial age: New perspectives on the science-technology relationship* (pp. 231–270).

Rip, A., Misa, T. J., & Schot, J. (Eds.). (1995). *Managing technology in society*. Pinter Publishers.

Schot, J. W. (1992). Constructive technology assessment and technology dynamics: The case of clean technologies. *Science, Technology, & Human Values, 17*(1), 36–56.

Schot, J. (1998). The usefulness of evolutionary models for explaining innovation. The case of the Netherlands in the nineteenth century. *History and Technology, an International Journal, 14*(3), 173–200.

Schot, J., & Geels, F. W. (2008). Strategic niche management and sustainable innovation journeys: theory, findings, research agenda, and policy. *Technology Analysis & Strategic Management, 20*(5), 537–554.

Schumpeter, J. A., & Nichol, A. J. (1934). Robinson's economics of imperfect competition. *Journal of Political Economy, 42*(2), 249–259.

Schuurman, D., De Marez, L., & Ballon, P. (2013). Open innovation processes in living lab innovation systems: Insights from the LeYLab. *Technology Innovation Management Review, 3*(11).

Spithoven, A., Frantzen, D., & Clarysse, B. (2010). Heterogeneous firm-level effects of knowledge exchanges on product innovation: Differences between dynamic and lagging product innovators. *Journal of Product Innovation Management, 27*(3), 362–381.

Urban, G. L., & Von Hippel, E. (1988). Lead user analyses for the development of new industrial products. *Management Science, 34*(5), 569–582.

Von Hippel, E. (1976). The dominant role of users in the scientific instrument innovation process. *Research Policy, 5*(3), 212–239.

Von Hippel, E., & Katz, R. (2002). Shifting innovation to users via toolkits. *Management Science, 48*(7), 821–833.

Andrew Paice is Head of the iHomeLab at Lucerne University of Applied Sciences. He earned a PhD in nonlinear systems and gathered 21 years of experience in industrial research and development before taking this position in 2018. The iHomeLab focuses in the areas of building intelligence with a focus on how humans will experience and interact with the buildings they live and work in. The aim is to make cutting edge research results useful in improving the quality of life for all people. Current applications are Smart Energy Management and Assisted Living for the elderly, applying technologies from automation, data science, artificial intelligence, and the Internet of Things.

Elena Malakhatka is currently conducting several studies on User Research applied to sustainable building operation and human-building interaction (HBI) at Chalmers Technical University at HSB living lab. Her doctoral project was done at KTH Live-in-Lab in "Data-driven service design for sustainable building operation and users' wellbeing."

Which Moonshot Metrics Matter?

William R. Cockayne and Tamara Carleton

Abstract Organizations that pursue the almost impossible—be it paradigm-shifting theories, scientific breakthroughs, technological invention, disruptive innovation, or game-changing products and services—are constantly grappling with the question, "What should we measure to gauge our potential opportunities, track our progress, or know we're investing correctly to deliver?" To understand this question, this chapter looks at rising opportunity for businesses to capture breakthroughs and inventions earlier in their development, the transition of the ARPA model (DARPA, I-ARPA, ARPA-E, and ARPA-H) from seeking future results to more near-term impact, and the internal measures used by Lenovo and Facebook to measure their progress. This chapter builds on the pioneering work done in Stanford University's Foresight program, a two-decade long effort that culminated in the launch of Stanford's Moonshots program and the book *Building Moonshots: 50+ Ways to turn Radical Ideas into Reality.*

1 Today's Science Fiction Is Tomorrow's Profits

As we approach the end of the first quarter of the twenty-first century, the world is pinning its future on the potential for unimaginable scientific breakthroughs, technological inventions, and game-changing products. While the pursuit of these goals would traditionally be confined to the technology press or niche industry sources, the need and opportunity for businesses to play a part in this change makes this headline news. A quick sample from business publications shows how these

W. R. Cockayne (✉)
Institute for Information Management, University of St. Gallen, St. Gallen, Switzerland
e-mail: william.cockayne@unisg.ch

T. Carleton
Department of Mechanical Engineering, Blekinge Institute of Technology, Karlskrona, Sweden
e-mail: tamara.carleton@bth.se

T. Carleton et al. (eds.), *Inventing the Almost Impossible*, Future of Business and Finance, https://doi.org/10.1007/978-3-031-36224-8_9

breakthroughs and inventions are already being measured using the language of traditional finance and innovation, as follows:

- Occidental makes a billion-dollar climate moonshot—so it can keep pumping oil (Morenne, 2023).
- Starship enterprise: the economics of Elon Musk's bold bet (Hollinger, 2023).
- Swedish "green steel" start-up plans €1.5bn fundraising (Levingston & Milne, 2023).

Moonshots, spaceships, and zero-carbon steel have only recently escaped science fiction, although they are already science fact, making the question of "What can we measure?" one that affects investors, entrepreneurs, engineers, and ultimately consumers.

Occidental Petroleum's largest shareholder is Warren Buffet, a famed value investor, who was also an early backer of electronic vehicle maker China-based BYD. Occidental is a successful oil and gas producer whose focus now includes nascent carbon capture technologies and bold metrics. While many companies and countries are touting 2050 as a time when they will see payback from today's belt-tightening or shifts in mindset, Occidental's history of making long-term investments makes this 27-year target one that is well within their second horizon of growth and new business development (Carleton & Cockayne, 2023). The company has set metrics for itself, which it is sharing with partners and investors, which includes building 135 carbon-dioxide removal plants by 2035, measurable goals which Occidental is already using to presell carbon credits to the likes of Airbus. Investors were quick to criticize this plan, as shareholder Kevin Holt of Invesco notes: "I think you have to be careful on the amount of capital you're [allocating] to any of these carbon-capture efforts. I prefer the moving-slowly path for anybody with a new technology, as opposed to just sprinting."

SpaceX is a fount of numbers, whether it concerns the Starship platform's ability to lift "150 metric tons of payload" into LEO (low-earth orbit), its promise of "launch prices well below $200" per kilogram, again into LEO, or the company's plan to add 7400 future satellites to the existing 3000 Starlink satellites already put into orbit. The ultimate metric of SpaceX is easy to understand; Occupy Mars proclaims the t-shirt that founder Elon Musk can often be seen wearing. During the initial SpaceX Starship launch, the company launched a recruiting discussion around their goal to reach "the moon, mars, and beyond." To date, the company has complemented this far-reaching vision with investments into technology development, engineering, and operations, building the stepping stones that the team needs to understand how to achieve each follow-on goal. It is within this model that even failures are considered measurable learning experiences. And at each step along this literal moonshot journey, SpaceX has been working closely with customers to align the company's solution with government and industry needs.

In February of 2021, H2 Green Steel (H2GS) announced its goal to create a "large-scale fossil-free steel plant in northern Sweden," with production starting as quickly as 4 years from that point. As proof of the endeavor's expected success, the leadership team pointed to founding organizations—SSAB, LKAB, and

Vattenfall—which explains the opportunity that H2GS had to borrow the playbook of nearby success, Northvolt. Tying together R&D, financing, and key customers from across the global steel industry, the team aligned with partner company Scania's need to transition the 5 tons of steel it needs to build a truck to zero carbon. In early 2023, the team was ready to raise 5 billion € and a commensurate level of debt financing. Using the results of early, never-before-scaled experiments, H2GS' timeline had slipped 2 years while helping to scope the launch as producing 2.5 million tons of steel for "about 10 terawatt hours of electricity annually." With an early delivery to Volvo Cars in Gothenburg, H2GS showed how to speed the commercialization of a seemingly impossible dream.

2 Commercializing the Impossible

Since the late 1950s, America's model of imagining, provoking, fostering, and funding the impossible has lived through the ARPA model (Bonvillian et al., 2020). This model is predicated on the creation of a vision that can ultimately "pay off" for the US government; "The real test of a good vision in R&D is whether others will commit resources to action, which will bring results in the future." The model has been proven right enough times—whether from the invention of the internet through stealth technology or GPS—to remain a pillar of America's pursuit of a currently unimaginable future.

The success of this model has led to the creation of similarly named groups, with aspirations for the same support of breakthrough outcomes (Tollefson, 2021). As I-ARPA (information) and ARPA-E (energy) reached their first decade in existence, the Biden administration funded ARPA-H (health) in early 2023. As Carleton (2020) writes, "however, the conundrum is that traditional R&D results may not be produced or easy to measure because the extent of far-reaching effects take time and are broadly distributed across society." As these new organizations emerge, are they adopting the same model of "paying others to invent the impossible"?

ARPA-E chose to directly address this conundrum. While the ARPA model broadly imagines "investing up front for an eventual, largely unknowable payoff," ARPA-E created their T2M, or Technology-to-Market, program to connect from the potentially unknowable, very-long view into a venture backed model of science-based or hard-technology-led entrepreneurship. ARPA-E's tech-to-market model aims to "help prepare teams to move their ground-breaking technical achievements out of the lab and towards real world impact in the energy sector" (ARPA-E, 2023). Without limiting the pursuit of moonshots across energy, ARPA-E's structure includes transformation as a core capability. While in previous eras a breakthrough invention fostered and funded by DARPA might be "put on the shelf" (George et al., 2018), such as the robotic surgery inventions developed at SRI during the advanced biomedical training (ABMT), including transformation in the defined model means talking with follow-on grant makers, early-stage investors, and supporting teams that know the goal is to move beyond mere breakthroughs and inventions to product, commercialization, and growth.

When ARPA-H was formed in March 2023, it focused on supporting "transformative biomedical and health breakthroughs—ranging from the molecular to the societal—to provide health solutions for all" (ARPA-H, 2023). In one of ARPA-H's first moves, it opened the opportunity for, in its words, "Accelerating Innovation through ARPA-H and FDA Collaboration," which focused on new incentives that could help create a bridge between the long range—"Health Science Futures" and the challenge to "Imagine the future of health"—with reimbursement for FDA activities that would help teams to develop and deliver new health innovations. As Carleton (2020) describes, this focus on connecting long-range technology vision with an organizational search for innovation opportunities is now a dominant feature of the ARPA model.

3 Building a Four Horizons Innovation Engine

Two companies that have publicly pioneered the development of four horizon models of innovation, anchoring to a bold vision and publishing their metrics for each interceding horizon, are Facebook and Lenovo.

Starting in 2016, then-Facebook CEO Mark Zuckerberg stood on stage of the company's F8 developer conference to explain the company's vision around Connectivity, AI, and AR/VR, which anchored the metrics that the company used to separate Ecosystems, Products, and Technologies (Cockayne, 2019a). Standing in front of a large horizons map, Zuckerberg described two metrics publicly that, until that point, were not well known outside the company.

Looking first at the third horizon named Technologies—often the stage of research and development—Zuckerberg explained that the transition to the second horizon of Products would occur when a product or service based on a new technology was being used by "one billion people." While this number seems in some ways unimaginable, for the Facebook team this was merely a measure of the popularity and adoption of any technology they chose to deliver to market. Beyond the actual number—which is significant in that it shows a seriousness in understanding when to transition an invention from R&D to growth and new business development—the existence of a metric that bridges that barrier shows a maturity in moving new ideas to commercialization that few companies possess. The classic retort from teams in research to "just throw it over the wall" or "that's a problem for the engineers to solve" is too often repeated to be merely a myth.

Zuckerberg proceeds to explain how Facebook measured the transition of a product into an ecosystem; coarsely, the company had a way to put advertisements in front of those billion users. Again, few companies can explain succinctly how a potential solution moves from development into growth, which is particularly true in many Silicon Valley venture-funded businesses where solutions are initially placed before customers at a measurable loss to the company's balance sheet (Barinka & Counts, 2023).

In 2021, Facebook was renamed as Meta, announcing a new fourth horizon of the metaverse. The company publicly announced that it would take 10 years and billions of dollars in investment before investors could expect positive results. In 2023, Meta returned to its 2016 roadmap to explain that the company had been hard at work on "AI" for almost a decade and that this would allow it to rapidly respond to the rise of LLM-based (large language models) AIs being integrated into products from Microsoft to startups. In public statements, Meta explained how it was driving the rebound of its advertising business using the years of AI investments it had been making, transitioning "AI" from the third horizon of Technologies to the first horizon of Ecosystems.

As another example, also in 2016, the executive team at China-based computing technology company Lenovo adopted a horizons model to better communicate with the English-speaking investment community (Cockayne, 2019b). In a letter to investors, Lenovo CEO Yang Yuanqing laid out the vision for the company, which imagined a time when Lenovo's investments in existing products, newly acquired business brands (notably Motorola), and as-yet-unnamed new inventions would place the company at the forefront of an envisioned "Smart Internet era."

In succeeding quarters, the CEO and his CFO led earnings calls that transitioned investors from thinking about the company just in terms of the last profit-and-loss (P&L) statement or mergers-and-acquisitions (M&A), which a cursory read of the business press makes apparent were the prior metrics by which Lenovo was measured. In the third quarter of 2019, a slide on new business growth from the Motorola acquisition laid claim to a headline that was traditionally one used by Apple in its iPhone marketing: "Innovation: Moto Z3 and 5G Mod announced as Verizon's first 5G solution."

While first to market with a product that integrated the latest in smartphone technologies would not be surprising for a company like Apple whose entire business was built around smartphone sales, for Lenovo this product was the result of an acquisition, in a product area that Lenovo had no US market experience, for a technology that was just emerging as a potential game-changer. This announcement pointed to a new way that Lenovo was measuring the success of its mergers and acquisitions efforts, new market entry, and partnerships.

In the same manner, a 2018 Lenovo earnings call touted success metrics for the Lenovo Star Wars: Jedi Challenges AR/VR kit. In a surprising move, Lenovo had released this immersive AR/VR product at a time when few companies were willing to take such a risk. Far from building the company's current sales or potential new markets, this product showcased Lenovo's ability to take what the market imagined was an impossible solution—a usable product that merged a cutting-edge head-mounted display—and delivered it directly to consumers. The metric that the company used for this solution was a wealth of positive reviews in the technology and consumer press, at a point where Google had pulled its early glasses from market, Magic Leap was struggling to deliver its head-mounted display to market, and Apple fans continued to sit on their hands in anticipation for just such a product.

4 Next Steps in Building Moonshots

Almost impossible initiatives might seem to thwart specific measurement, even while investors and senior management push to see the market opportunity and business case presented in hard numbers. However, there are different ways to use numbers that allow radical visions, which seem intractable, to become closer to reality for partners, investors, and others. This chapter describes multiple metrics used publicly by Occidental Petroleum, SpaceX, H2 Green Steel, Meta, and Lenovo as initial evidence "in the wild" for quantifying the scope and progress of building moonshots.

Going forward, this chapter raises the need to develop a global Moonshots Index that aims to track, collate, distill, and visualize the spectrum of metrics used across the four horizons, especially for horizon four into horizon three when ambiguity is the highest. In doing so, a subset of global leaders can be identified. Another benefit of a Moonshot Index will be to track performance indicators across multiple categories, offering a data-driven and unbiased resource for those seeking benchmarks for their own game-changing innovation.

References

ARPA-E. (2023). *ARPA-E tech-to-market advisors webpage*. https://arpa-e.energy.gov/career/arpa-e-tech-market-advisors

ARPA-H. (2023). *About ARPA-H*. https://arpa-h.gov/

Barinka, A., & Counts, A. (2023, April 26). Meta shares surge as digital ad rebound fuels AI investments. *Bloomberg*. https://www.bloomberg.com/news/articles/2023-04-26/meta-revenue-beats-estimates-on-facebook-advertising-recovery

Bonvillian, W. B., Van Atta, R., & Windham, P. (Eds.). (2020). *The DARPA model for transformative technologies: Perspectives on the U.S. Defense Advanced Research Projects Agency*. Open Book Publishers.

Carleton, T. (2020). The value of vision in radical technological innovation. In W. B. Bonvillian, R. Van Atta, & P. Windham (Eds.), *The DARPA model for transformative technologies: Perspectives on the U.S. Defense Advanced Research Projects Agency*. Open Book Publishers.

Carleton, T. L., & Cockayne, W. R. (2023). *Building moonshots: 50+ ways to turn radical ideas into reality*. John Wiley & Sons.

Cockayne, W. (2019a). *Facebook's 10-year product roadmap*. Innovation Leadership Group Case.

Cockayne, W. (2019b). *Measuring Lenovo's innovation engine*. Innovation Leadership Group Case.

George, E. I., Brand, T. C., LaPorta, A., Marescaux, J., & Satava, R. M. (2018). Origins of robotic surgery: From skepticism to standard of care. *JSLS: Journal of the Society of Laparoendoscopic Surgeons, 22*(4), e2018.00039. https://doi.org/10.4293/JSLS.2018.00039

H2 Green Steel. (2021, February 23). *H2 Green Steel to build large-scale fossil-free steel plant in northern Sweden* [Press release]. https://www.h2greensteel.com/latestnews/bibendum-sit-malesuada

Hollinger, P. (2023, April 26). Starship enterprise: The economics of Elon Musk's bold bet. *Financial Times*. https://www.ft.com/content/d6147eae-3493-4b70-bc5d-41610a680ebb

Levingston, I., & Milne, R. (2023, April 24). Swedish 'green steel' start-up plans €1.5bn fundraising. *Financial Times*. https://www.ft.com/content/e9f2fc05-87ed-4ece-bce4-ca79e0e4b21c

Morenne, B. (2023, April 10). Occidental makes a billion-dollar climate moonshot—so it can keep pumping oil. *Wall Street Journal.* https://www.wsj.com/articles/occidental-plans-to-suck-carbon-from-the-airso-it-can-keep-pumping-oil-2990c5a
Tollefson, J. (2021). The rise of 'ARPA-everything' and what it means for science. *Nature, 595*, 483–484.

William R. Cockayne has spent his life building great teams who can imagine, invent, and deliver the future. He is a visionary technologist with a passion for understanding the latest technologies. He holds a doctorate in mechanical engineering from Stanford University and is the co-author of *Building Moonshots: 50+ Ways to Turn Radical Ideas into Reality* and the *Playbook for Strategic Foresight and Innovation.*

Tamara Carleton is a globally recognized expert in radical innovation. As founder and CEO of Innovation Leadership Group, Tamara works with business and government leadership teams to help them build innovation capability and lay out bold futures. With a penchant for creating new methods and tools, Tamara is also a renowned educator, teaching strategic foresight and innovation at multiple universities worldwide. She holds a doctorate in mechanical engineering from Stanford University, where she explored DARPA's enduring innovation practices. She is the author of *Building Moonshots: 50+ Ways to Turn Radical Ideas into Reality* and the *Playbook for Strategic Foresight and Innovation.*

MIX
Papier aus verantwortungsvollen Quellen
Paper from responsible sources
FSC® C105338

FSC
www.fsc.org

If you have any concerns about our products,
you can contact us on
ProductSafety@springernature.com

In case Publisher is established outside the EU,
the EU authorized representative is:
Springer Nature Customer Service Center GmbH
Europaplatz 3, 69115 Heidelberg, Germany

Printed by Libri Plureos GmbH
in Hamburg, Germany